Lectures on Special Relativity

Related Pergamon Titles of Interest

Books

ATWATER
Introduction to General Relativity

BOWLER
Lectures on Statistical Mechanics

GINZBURG
Theoretical Physics & Astrophysics (A Selection of Key Problems)

LANDAU & LIFSHITZ
Course of Theoretical Physics (in 10 volumes)

PATHRIA
The Theory of Relativity, 2nd edition

SERRA *et al.*
Introduction to the Physics of Complex Systems (The mesoscopic approach to fluctuations, non linearity and self-organization)

TORRETTI
Relativity & Geometry

Journals

Journal of the Franklin Institute

Reports on Mathematical Physics

USSR Computational Mathematics & Mathematical Physics

Full details of all Pergamon publications/free specimen copy of any Pergamon journal available on request from your nearest Pergamon office.

Lectures on Special Relativity

by

M. G. BOWLER
Department of Nuclear Physics,
Oxford University

LIBRARY
CHRISTIAN BROTHERS COLLEGE
650 E. PARKWAY S.
MEMPHIS. TENN. 38104
80067

PERGAMON PRESS
OXFORD · NEW YORK · BEIJING · FRANKFURT
SÃO PAULO · SYDNEY · TOKYO · TORONTO

U.K.	Pergamon Press, Headington Hill Hall, Oxford OX3 0BW, England
U.S.A.	Pergamon Press, Maxwell House, Fairview Park, Elmsford, New York 10523, U.S.A.
PEOPLE'S REPUBLIC OF CHINA	Pergamon Press, Qianmen Hotel, Beijing, People's Republic of China
FEDERAL REPUBLIC OF GERMANY	Pergamon Press, Hammerweg 6, D-6242 Kronberg, Federal Republic of Germany
BRAZIL	Pergamon Editora, Rua Eça de Queiros, 346, CEP 04011, São Paulo, Brazil
AUSTRALIA	Pergamon Press Australia, P.O. Box 544, Potts Point, N.S.W. 2011, Australia
JAPAN	Pergamon Press, 8th Floor, Matsuoka Central Building, 1-7-1 Nishishinjuku, Shinjuku-ku, Tokyo 160, Japan
CANADA	Pergamon Press Canada, Suite 104, 150 Consumers Road, Willowdale, Ontario M2J 1P9, Canada

Copyright © 1986 M. G. Bowler

All Rights Reserved. No part of this publication may be reproduced, stored in a retrieval system or transmitted in any form or by any means: electronic, electrostatic, magnetic tape, mechanical, photocopying, recording or otherwise, without permission in writing from the publishers.

First edition 1986

Library of Congress Cataloging in Publication Data

Bowler, M. G.
Lectures on special relativity.
Includes index.
1. Special relativity (Physics) I. Title.
QC173.65.B69 1986 530.1'1 86-9463

British Library Cataloguing in Publication Data

Bowler, M. G.
Lectures on special relativity.
1. Relativity (Physics)
I. Title
530.1'1 QC173.65

ISBN 0-08-033939-5 Hard cover
ISBN 0-08-033938-7 Flexicover

Printed in Great Britain by A. Wheaton & Co. Ltd., Exeter

PREFACE

"The special relativity theory, which was simply a systematic extension of the electrodynamics of Maxwell and Lorentz, had consequences which reached beyond itself ..."

Einstein, writing in *The Times*, November 28 1919.

Imagine a large number of starships, containing well found laboratories and competent physicists, moving along different straight lines at different constant speeds. Experiments can be carried out on board each ship, and astrophysical observations may be made from each ship. In each and every ship scientific instruments work in the same way, standard laboratory experiments give the same results and from both laboratory experiments and astrophysical observations everyone infers the same fundamental laws of physics ... the swarm of starships exists in a universe governed by a principle of relativity.

Look at some individual experiments. Interstellar analogues of Galileo and Newton play with billiard balls in their laboratories and find them to be governed by Newtonian mechanics. Analogues of Huyghens, Fresnel and Young study optics, analogues of Coulomb, Ampere and Faraday study electricity and magnetism. Analogue Maxwells synthesise the Maxwell equations. In every ship the same physical laws are found, and in particular every analogue Michelson discovers that the speed of light is independent of its direction, and it has the same measured value in every ship. Now there is a problem, because the invariance of the speed of light is at odds with the vector addition of velocities worked out by the mechanics.

The universe is so constructed that the speed of light is indeed measured to be the same, in all directions, regardless of which starship the measurement is made in. The vector addition of velocities worked out using billiard balls or relatively low velocity excursion modules is in fact wrong, and measured velocities do not compound in that way as any velocity becomes significant in comparison with that of light. This is all experimental fact and at first sight seems incomprehensible.

Incomprehension lifts (to some extent) with the realisation that measurements are made with instruments, instruments are made of matter, and matter is complicated ... so complicated that a measuring stick moving very fast relative to any one of our starships is contracted along the line of relative motion, a clock moving very fast runs slow, and mass is an increasing function of velocity. All these features are inherent in classical electromagnetism, and must be shared by all forms of matter in a universe governed by a principle of relativity. Such a principle places stringent restrictions on the form of physical law and this is the matter of the special theory of relativity.

Special relativity is now 80 years old and is diffusing into the early stages of university physics courses. There is a feeling abroad that exciting things like relativity should be taught as early as possible, the implication being that the physics of the 19th century is not exciting. This is unfortunate, for the physics of the 19th century is enormously impressive and exciting and much of it is novel to the student commencing university work. Special relativity only just avoided the 19th century and is the daughter of classical electromagnetic theory, which was formulated only some 40 years earlier.

An early introduction to special relativity has disadvantages. One is that for lack of appropriate mathematical tools rather arid problems in kinematics assume a disproportionate importance. Another is that familiarity breeds contempt, and many will conclude that relativity, too, is dull. But to my mind the greatest disadvantage is that an early course in relativity usually precedes study of electromagnetism and the student is thereby deprived of a background which makes sense of the apparently nonsensical.

A number of years ago I lectured to second year undergraduates at Oxford on the subject of special relativity, but it was only recently that I was goaded (by the perversity of some of my colleagues) to write up that material for publication. My aim when preparing the lectures which appear in this book was to give a concise account of the essential content of special relativity, without compromising the development of the subject by avoiding (relatively) advanced mathematics. Since the lectures were designed for students who knew little or no relativity, I was also concerned to pay particular attention to those difficulties, conceptual rather than mathematical, which invariably snare the vast majority.

Lecture 1 is concerned with the nature of a principle of relativity, the definition of inertial frames and the relationship between coordinates measured in different frames. The conflict between Newtonian mechanics, electromagnetism and a universal principle of relativity is spelt out and illustrated, and the lecture ends with a demonstration that electromagnetic matter will suffer Lorentz contraction. I believe that the realisation that classical electromagnetism predicts that electromagnetic measuring rods are shrunk and electromagnetic clocks are slowed when moving relative to the putative aether is of great help in understanding that the Lorentz transformations are physically perfectly acceptable. Lecture 2 is devoted to a careful and rigorous derivation of the Lorentz transformations themselves, with particular reference to their reciprocal nature. Lecture 3 deals with the related phenomena of time dilation and Lorentz contraction. Here I have been concerned to make clear which way round the formulae should be worked, the reciprocal nature of each phenomenon, and the absence of paradox.

Lecture 4 is largely devoted to the mathematical techniques which are necessary for grasping the essence of special relativity and which make so much easier many calculations where they cannot be held essential. These techniques are then put to work in obtaining the relativistic Doppler shift and in studying the covariance of Maxwell's equations. The requirement that conservation of energy and momentum be covariant leads in Lecture 5 to the identification of the correct expressions for energy and momentum as components of a 4-vector. The famous relation $E = mc^2$ is extracted and its physical significance is discussed. 4-forces are introduced and the relations between force, rate of change of momentum and acceleration are developed. The whole works is illustrated by writing the Lorentz force law in manifestly covariant terms.

Special relativity is verified continually in any high energy physics laboratory and the manipulation of relativistic kinematics is a tool of the trade of the high energy physicist. Lecture 6 is concerned wholly with the tricks of this trade and is liberally illustrated with real examples and problems drawn from high energy physics. It is for the reader who wishes to become familiar with this practical application of relativity and may be skimmed or skipped by those who find such application repellent.

Lecture 7 returns to the subject of the correct equations of motion of particles experiencing a force and deals briefly with the treatment of physical laws obtaining in an accelerated frame of reference—one of our starships when the drive is on. The interminable problem known as the twin paradox is then treated in considerable detail—80 years after the genesis of special relativity this problem continues to perplex successive generations of students, and not a few of their seniors.

Lecture 8 is to a large extent disjoint from the rest of the book. In the context of special relativity, neither matter, energy nor information can be propagated faster than light. But there are lots of things which go faster than light and in the absence of a careful analysis the consistency of special relativity as a covering principle for the physical world may be questioned. The lecture consists of a number of examples of things which go faster than light, ranging from the homely example of scissors through the old problems of phase and group velocity in classical physics to examples drawn from astrophysics, such as the apparent superluminal expansion of certain quasars. The conceptual problems encountered in arranging a marriage between relativity and quantum mechanics are discussed.

It would be best if the reader of this book were already acquainted with electromagnetism up to the level of Maxwell's equations and waves in empty space. I hope however that the book is sufficiently self explanatory that those whose studies are not that far advanced will nonetheless be able to acquire an understanding of the principle of relativity and the marvellous construction of the physical world which is expressed therein.

As in my previous lecture note volume, *Lectures on Statistical Mechanics* (Pergamon, 1982), PROBLEMS are scattered liberally throughout the text. They should be regarded as an integral part of the course. Many are simple exercises designed to further understanding of the fundamental material, while others are there to inculcate facility in solving realistic problems. The vast majority are very quick and easy ... but not all.

CONTENTS

Lecture 1 WHAT IS RELATIVITY?

A particular physical system satisfies a principle of relativity if the laws governing that system take the same form in all frames of reference moving with constant velocity. Thus formulated, relativity is a property of a particular set of physical theories. If all laws of physics take the same form, then the principle of relativity becomes rather a conceptual framework which must be satisfied by any particular theory, and a fundamental property of the physical universe.

Special relativity (Einstein's relativity) provides such a framework, into which the whole of physics seems to fit and which provides general principles within which any new theoretical model should be constructed. This framework has superseded that of Galileo and Newton (which was very fruitful and is very accurate provided only relatively small velocities are involved), which in turn superseded that of Aristotle (essentially useless). Relativistic mechanics (the theory of mechanics which satisfies the principles contained within special relativity) is no more difficult than Newtonian mechanics: electrodynamics as formulated by Maxwell satisfied these principles from the outset.

Consider a physicist studying a particular physical system. The simplest system he could study is a single particle not acted on by any force. It is an abstraction from experience that such a particle moves in a straight line at constant speed, when measured with linearly calibrated instruments, unless the physicist and his instruments are themselves accelerated. A frame of reference in which such a free test particle moves uniformly is called an *inertial frame* of reference.

A second observer, moving with his laboratory paraphenalia with constant velocity relative to the first, will see the single free test particle move with a different constant velocity. (The simplest situation is a particle at rest with respect to the apparatus of one observer, but moving with constant velocity with respect to the other). These considerations serve to define the set of inertial frames of reference.

We now let our observers watch the behaviour of a more complicated physical system: billiard balls on elastic strings, an atom emitting radiation, a star cluster, or whatever you like. If a universal principle of relativity holds, any physicist in any inertial frame will deduce the same laws of physics: the same in form and the same in numerical content; that is, the laws of physics assume the same form in all inertial frames of reference.

However philosophically attractive such a principle may seem, its applicability to the real world must be tested by investigating the real world (and remember that in accelerated frames of reference the laws of physics seem to assume a different form).

Newtonian mechanics embodies a principle of relativity — not Einstein's relativity but Galileo's.

Consider Newton's laws:

1. 'A body will move in a straight line at constant speed so long as no external force acts on it'.

(This is not circular: it is implicit that something must be around to produce a force).

We had better add explicitly the qualification 'observed from an inertial frame, measurements being made with linearly calibrated instruments' and assume that we really can make such measurements. This law may be taken as defining the family of inertial frames.

2. Force = mass × acceleration

which may be expressed in any of the forms

$$\mathbf{F} = m\mathbf{a} \qquad \text{or} \qquad \mathbf{F} = m\frac{d^2\mathbf{x}}{dt^2} \qquad \text{or} \qquad F_i = m\frac{d^2x_i}{dt^2}$$

F and **a** (and **x**) are vectors: this equation retains its form and numerical content under both rotations and translations of the coordinate system. The quantities x_i are the three orthogonal spatial coordinates and t (time) is a universal parameter which can be eliminated to yield the trajectory (as opposed to the equation of motion) of an object (for example the parabola described by a flung brick or the ellipse traced by a planet).

How are the coordinates of a particle in one frame related to the coordinates of the same particle as measured in another (inertial) frame? Let the two frames move with relative velocity v along common x axes

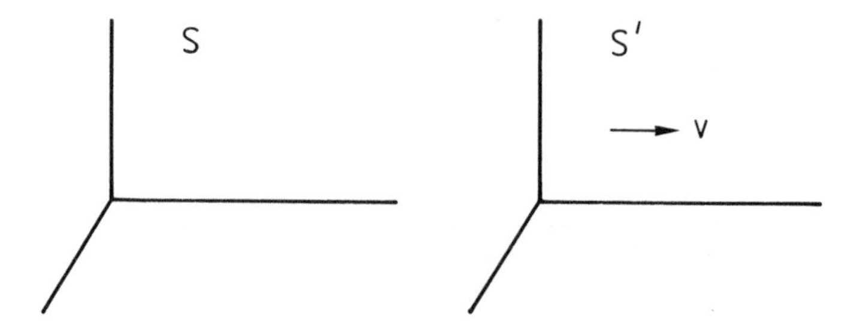

A marker at fixed x' in S' moves with velocity v along the x axis in S, such that

$$\left.\frac{dx}{dt}\right|_{S'} = v \tag{1.1}$$

and conversely

$$\left.\frac{dx'}{dt'}\right|_{S} = -v \tag{1.2}$$

These relations constitute the definition of relative velocity. The Galilean relation between x and x' is (choosing coincident origins for convenience)

$$\begin{aligned} x' &= x - vt \qquad t' = t \qquad \text{or} \\ x &= x' + vt' \qquad t = t' \end{aligned} \tag{1.3}$$

These are consistent equations which satisfy (1.1) and (1.2), but they are NOT the ONLY equations satisfying (1.1) and (1.2).

[PROBLEM: Find another set]

Suppose that in S'

$$F'_{x'} = m\frac{d^2x'}{dt'^2}$$

The Galilean transformation then yields

$$\frac{dx'}{dt'} = \frac{dx'}{dt} = \frac{dx}{dt} - v \qquad \text{or} \qquad u'_{x'} = u_x - v$$

$$\frac{d^2x'}{dt'^2} = \frac{d^2x}{dt^2}$$

and so if $\mathbf{F}$ and m are the same in all such inertial frames the Galilean transformation takes you from Newton's laws in one frame to Newton's laws in another — same form and same numerical content. Newton's laws are *covariant* with respect to the Galilean transformations. Look into the origin of force a little further and introduce a potential:

$$\frac{\partial}{\partial x'} = \frac{\partial}{\partial x} \qquad \text{so} \qquad \nabla'\phi' = \nabla\phi$$

and

$$m\frac{d^2\mathbf{x}'}{dt'^2} = -\nabla'\phi' \qquad \text{transforms into}$$

$$m\frac{d^2\mathbf{x}}{dt^2} = -\nabla\phi$$

and the equation holds equally well for all observers in inertial frames, provided that the scalar potential ϕ has the same value at the location of the particle in all frames; that is, ϕ is invariant under the Galilean transformations.

The corollary of all this is that there is no way of using Newtonian mechanics to define a meaningful absolute velocity — but if this seems obvious, contrast the case of accelerations.

The relation

$$\mathbf{u}' = \mathbf{u} - \mathbf{v}$$

may seem obvious. It isn't. Many people have terrible difficulties with problems involving vector addition of velocities at the first encounter. Obvious or not, it works — at low velocities.

Let's go on to another part of antediluvian physics — optics and electromagnetism as amalgamated by Maxwell. Maxwell's equations give the velocity of light equal to c (see Appendix) and we at once ask: with respect to what?

Suppose we suggest half an answer: with respect to some particular inertial frame (the aether frame). Then we expect that in some other inertial frame that a pulse of light would propagate with velocity $\mathbf{c}' = \mathbf{c} - \mathbf{v}$ and Maxwell's wonderful equations would not be true in any other frame. Why worry? After all, the wave equation for sound gives the velocity with respect to the medium supporting it and this is Newtonian physics. Surely we can keep a principle of relativity in the same sense for electromagnetism? Philosophically yes, but such a principle would be barren for light pervades the entire universe and would seem to define a universal fancy reference frame defined by Maxwell's equations being true in that frame only.

Let's look at the troubles with light a bit more quantitatively. The wave

$$\mathbf{E} = \mathbf{E}_0 \sin(\mathbf{k}.\mathbf{x} - \omega t) \qquad k^2 = \frac{\omega^2}{c^2} \tag{1.4}$$

and its associated magnetic field satisfy Maxwell's equations in empty space. The argument of the oscillatory function is the phase of the wave and is a pure number (proportional to a number of wave crests). It may be visualised as the number of wave crests passing the point $\mathbf{x}$ (or $\mathbf{x}'$) between the time of arrival of the crest which was at $\mathbf{x} = 0$, $t = 0$ and the time t, multiplied by (-2π). Since the origins of S, S' were defined to coincide and $\mathbf{x}'$ at $t' = t$ refers to the same point in space as $\mathbf{x}$ at the time t, the counting operation will yield the same number in any inertial frame and the phase

$$\phi = \mathbf{k}.\mathbf{x} - \omega t$$

will thus be an invariant.

[PROBLEM: Convince yourself of this. Remember that a wave crest passing the origin could be marked by introducing a small distortion and the time t could be signalled at $\mathbf{x}$ by running up a flag]

Rotating coordinates so that $\mathbf{x}$ lies along $\mathbf{k}$

$$\phi = kx - \omega t$$

and

$$\left.\frac{dx}{dt}\right|_{\phi} = \frac{\omega}{k} = c \tag{1.5}$$

and is the phase velocity in the direction of $\mathbf{k}$, normal to the wavefront. (Notice that the velocity with which the intersection of a plane of constant phase with any other axis travels is >c).

Then observer O (in the frame in which Maxwell's equations hold) measures a phase ϕ at x, t, and propagation vector $\mathbf{k}$, frequency ω. An observer O' in some other frame measures the same value of the phase ϕ but $\mathbf{k}'$, ω' such that

$$\mathbf{k}'.\mathbf{x}' - \omega' t' \equiv \mathbf{k}.\mathbf{x} - \omega t$$

where the relation is an identity because the phase is the same number in all frames for any values of x, t. Using the Galilean relations $x' = x - vt$, $t' = t$ and equating coefficients we have

$$\mathbf{k}'.(\mathbf{x} - \mathbf{v}t) - \omega' t \equiv \mathbf{k}.\mathbf{x} - \omega t$$

and hence

$$\mathbf{k}' = \mathbf{k}, \quad \omega' = \omega - \mathbf{v}.\mathbf{k}' = \omega - \mathbf{v}.\mathbf{k} = \omega(1 - \frac{\mathbf{v}.\hat{\mathbf{k}}}{c}) \tag{1.6}$$

where $\hat{\mathbf{k}}$ is a unit vector normal to the wave front. This of course is the familiar Doppler shift. The phase velocity in S' is given by

$$c' = \frac{\omega'}{k'} = c - \mathbf{v}.\hat{\mathbf{k}} \qquad \text{(not equal to c)} \tag{1.7}$$

Note that c' is the phase velocity, normal to the wavefront and is not in general equal to the velocity of a pulse of light (or a photon) given by $\mathbf{u}' = \mathbf{c} - \mathbf{v}$ (although $\hat{\mathbf{k}}.\mathbf{u}' = c'$). In S' a pulse of radiation does not propagate normal to the wavefront (and it does if Maxwell's equations are true).

For propagation along the common x axis,

$$\sin(k'x' - \omega' t') = \sin(kx' - \omega(1 - \frac{v}{c})t')$$

and as $v \to c$ this equation predicts that an observer chasing the electromagnetic wave sees a field oscillating sinusoidally in space but not in time. This is not a solution of Maxwell's equations: the free field equations are not covariant with respect to the Galilean transformations.

Here is another example, in which sources of the fields are important. Consider the force per unit length acting between two (infinitely long) strings of charge, at rest in a frame where Maxwell's equations hold:

This is an elementary problem in electrostatics. The relevant Maxwell equation is

$$\nabla.\mathbf{E} = 4\pi\rho \qquad \text{where } \rho \text{ is (three dimensional) charge density}$$

$$\text{whence} \qquad \int \mathbf{E}.d\mathbf{S} = 4\pi q \qquad \text{where } q \text{ is the charge enclosed}$$

within the surface S. If the charge per unit length on a string is σ, then $\mathbf{E}$ is radial and given by

$$2\pi E(d) = 4\pi\sigma$$

$$E(d) = \frac{2\sigma}{d} \tag{1.8}$$

The force per unit length is

$$F_E = \frac{2\sigma^2}{d} \qquad \textit{outward} \tag{1.9}$$

Now suppose the strings move in the direction of their lengths with velocity v relative to a frame in which Maxwell's equations hold. The electrical force still takes the form (1.9), but there is also a magnetic force between the two strings. For steady currents

$$\nabla \times \mathbf{B} = \frac{4\pi}{c}\mathbf{J}$$

and applying Stokes' theorem

$$2\pi d B(d) = \frac{4\pi}{c} I$$

$$B(d) = \frac{2I}{cd} \tag{1.10}$$

The force is $IB(d)/c$ per unit length, or

$$F_B = \frac{2I^2}{c^2 d} \qquad \textit{inward} \tag{1.11}$$

and I (charge per second) is given by

$$I = \sigma v$$

so that the total force per unit length is

$$F = \frac{2\sigma^2}{d}\left(1 - \frac{v^2}{c^2}\right) \tag{1.12}$$

[The two infinitely long strings were chosen to get rid of retardation effects. The results are correct in a frame in which Maxwell's equations are true, although the result for the electric field of a moving string may surprise you if you know the result for the transverse field of a moving (point like) particle]. If the measured value of σ is the same in the Maxwell frame and in a frame moving with the strings, then the two different values obtained for F are inconsistent with Galilean relativity and again the appropriate Maxwell equations could not hold in the moving frame.

Since the validity of a universal principle of relativity is to be determined experimentally, consider the results of experiments which tested explicitly for the existence of a preferred frame. The original experiment was that of Michelson and Morley, in which a Michelson interferometer with two equal length arms at right angles was rotated through 90° while the fringes formed were observed. A movement of the fringes by an amount proportional to $\frac{v^2}{c^2}$, where v is the velocity of the interferometer through the aether (or with respect to the Maxwell frame) was expected: no fringe shift was observed.

[PROBLEM: Work out the expression for the fringe shift as a function of the arm lengths, wavelength of light employed and velocity v. Evaluate it for arm lengths of 1m and $v \sim 30\,\mathrm{km\,s^{-1}}$ (orbital velocity of the earth), $v \sim 200\,\mathrm{km\,s^{-1}}$ (velocity of the sun around the galaxy) and $v \sim 400\,\mathrm{km\,s^{-1}}$ (velocity of the sun with respect to the microwave background).]

The most sensitive experiment looking for motion relative to a preferred frame employed the Mössbauer effect to look for Doppler shift variations, eq. (1.6). 14.4 KeV photons from the decay of ^{57}Fe (fed by β emission from ^{57}Co) have a natural line width $\frac{\Delta\omega}{\omega} \sim 10^{-13}$ and at room temperature the iron nuclei are sufficiently locked into the crystal lattice that the recoil momentum is taken up by a patch of crystal rather than by the daughter nucleus. First order Doppler shifts due to thermal jiggling are also absent. If an absorbing iron foil (enriched in ^{57}Fe) is moved towards or away from a 14.4 KeV ^{57}Fe source, the change in absorbtion as a function of velocity is sensitive to velocities $\ll 0.1\,\mathrm{mm\,s^{-1}}$.

Consider a source moving through the preferred frame with velocity $\mathbf{v}_s$, frequency (in the source frame) ω_s. Then from (1.6)

$$\omega_s = \omega\left(1 - \frac{\mathbf{v}_s}{c}.\hat{\mathbf{k}}\right)$$

where ω is the frequency in the Maxwell frame. The frequency measured in the absorber frame is similarly

$$\omega_a = \omega\left(1 - \frac{\mathbf{v}_a}{c}.\hat{\mathbf{k}}\right)$$

so that

$$\omega_s - \omega_a = \omega\frac{\mathbf{v}_a - \mathbf{v}_s}{c}.\hat{\mathbf{k}}$$

$\mathbf{v}_s = \mathbf{u}_s - \mathbf{v}$, $\mathbf{v}_a = \mathbf{u}_a - \mathbf{v}$ where $\mathbf{u}_s$, $\mathbf{u}_a$ are the laboratory velocities of the source and absorber and $\mathbf{v}$ is the velocity of the laboratory through the preferred frame. Thus

$$\omega_s - \omega_a = \omega\left(\frac{\mathbf{u}_a - \mathbf{u}_s}{c}\right).\hat{\mathbf{k}} \tag{1.13}$$

At first sight, (1.13) is independent of $\mathbf{v}$. This is not in fact the case, for the scalar product contains dependence on $\mathbf{v}$. The direction of travel of the photons $\hat{\mathbf{p}}$ is offset from $\hat{\mathbf{k}}$ by $\sim \frac{\mathbf{v}}{c}$ and so if $\hat{\mathbf{p}}.(\mathbf{u}_a - \mathbf{u}_s) = 0$ then

$$\omega_s - \omega_a = \omega(\mathbf{u}_a - \mathbf{u}_s).\frac{\mathbf{v}}{c^2} \tag{1.14}$$

In the aether drift experiment, the source and absorber were mounted at opposite ends of the diameter of a turntable which could be rotated at high speed. Then $\hat{\mathbf{p}}.(\mathbf{u}_a - \mathbf{u}_s)$ is indeed zero and so $\frac{\omega_s - \omega_a}{\omega}$ will oscillate with the rotational frequency of the turntable and an amplitude proportional to the component of $\mathbf{v}$ in the plane of the turntable. A limit $v < 5\,\mathrm{cm\,s^{-1}}$ has been set [Reference G. R. Isaak, *Phys. Bull.* **21** 255 (1970)].

[PROBLEM: (There is no easy way to solve this—it has to be worked out carefully and requires some playing around. These sort of considerations are of enormous importance in real experimental physics). At first sight, $\hat{\mathbf{p}}.(\mathbf{u}_a - \mathbf{u}_s) = 0$ seems obvious for the turntable — the photon travels along a diameter, doesn't it? The answer is no, because of the finite speed of light in the laboratory, which changes with orientation of the source and absorber relative to $\mathbf{v}$. So first show that $\hat{\mathbf{p}}.(\mathbf{u}_a - \mathbf{u}_s)$ really is zero, for source and absorber opposite, each at a radius R, independent of the velocity with which the photon travels in the laboratory. But you will never locate them precisely at a distance R, so take radii R_s and R_a and show that $\hat{\mathbf{p}}.(\mathbf{u}_a - \mathbf{u}_s)$ is still zero. Then remember that the source and absorber will not be precisely located on a diameter (and in any case cover finite areas) so show that even so $\hat{\mathbf{p}}.(\mathbf{u}_a - \mathbf{u}_s)$ is zero!]

So experiments designed to detect the motion of a terrestrial laboratory relative to an aether frame reveal no such effect — what do we do? Make some guesses (and in the early stages we don't worry if the philosophers tell us our hypotheses are *ad hoc*.)

1. Perhaps an aether (with mechanical properties) is dragged along as a boundary layer with the earth? NO (because the apparent angular position of stars changes with a period of 1 year — stellar parallax.)

2. Perhaps the preferred frame is always the emitter frame? NO (initially ruled out by observation of binary stars, but this evidence is probably vitiated by the phenomenon of extinction — light is absorbed and reradiated by the interstellar medium and the memory of source velocity would disappear in a few light years. But the velocity of photons from the decay of $\pi^{0\prime}s$ travelling with a velocity $\sim c$ in the laboratory has been measured and is $\sim c$ rather than $2c$ [T. Alvager et al., *Phys. Lett.* **12** 260 (1964)])

3. Given a theory with a preferred frame, do we really expect to MEASURE a direction dependent velocity?

Lorentz's answer to this question was NO, because Maxwell's equations, taken as true in one frame, lead us to expect a longitudinal contraction of apparatus moving with respect to this frame (and so the Lorentz contraction is more than an *ad hoc* hypothesis).

Suppose a measuring stick (or the arm of a Michelson interferometer) consists of charges sitting at the bottom of electrostatic potential wells. In regions of space where the charge density is zero, the scalar potential may be chosen to be a solution of the equation

$$\nabla^2\phi - \frac{1}{c^2}\frac{\partial^2\phi}{\partial t^2} = 0 \tag{1.15}$$

in the Maxwell frame.

The scalar potential in a measuring stick at rest with respect to the aether thus satisfies the Laplace equation

$$\nabla^2\phi = 0$$

In motion, the scalar potential ϕ_M of the stick must satisfy (1.15) in the Maxwell frame, because the potential at fixed x changes with time. If the stick moves in the x direction with velocity v, then

$$\phi_M = \phi_M(x \pm vt)$$

and so

$$\frac{\partial^2 \phi_M}{\partial t^2} = v^2 \frac{\partial^2 \phi_M}{\partial x^2}$$

Freezing the picture at a given instant, the potential must satisfy the equation

$$\frac{\partial^2 \phi_M}{\partial x^2}\left(1 - \frac{v^2}{c^2}\right) + \frac{\partial^2 \phi_M}{\partial y^2} + \frac{\partial^2 \phi_M}{\partial z^2} = 0 \tag{1.16}$$

this pattern moving with velocity v through the aether. This equation is just Laplace's equation again, in scaled coordinates

$$\nabla_M^2 \phi_M = 0$$

where

$$x_M = \frac{x}{\sqrt{1 - v^2/c^2}}, \quad y_M = y, \quad z_M = z.$$

We may therefore set $\phi_M(\mathbf{x}_M) = \text{const}\ \phi(\mathbf{x})$ and the distance between maxima and minima of ϕ_M is the same, measured in terms of x_M, as the separation between maxima and minima of ϕ, measured in terms of the true Maxwell frame coordinate x. Thus for the moving rod, the distance between maxima and minima measured in terms of x is reduced by a factor $\sqrt{1 - v^2/c^2}$. This is precisely what is needed to account for the null result of the Michelson-Morley experiment with equal arm lengths. This result suggests that charge density is likely to be increased (in order to match the equipotentials) and if for a moving string of charge $\sigma_M = \frac{\sigma}{\sqrt{1-v^2/c^2}}$ THEN we no longer have any problem with the (transverse) force per unit length acting between strings of charge (eq. (1.12)). [This also removes the apparent contradiction between the transverse fields of a moving point particle and eq. (1.8)].

This is of course not the whole story. A purely static distribution of charges is not stable, but there are a number of other interesting effects which are relevant. The momentum in the electromagnetic field of a point (or very small) particle can be calculated via Maxwell's equations and it rises faster than linearly with velocity in the Maxwell frame, giving rise to different transverse and longitudinal masses. The shape of the orbit of a bound state of two charges moving through the Maxwell frame can be calculated and indeed is shrunk by a factor $\sqrt{1 - v^2/c^2}$ (where v is the velocity of the centre of mass of the pair through the aether) AND the period is increased, relative to an identical molecule at rest. Setting aside the problems of constructing a consistent classical theory of electromagnetic matter, there is every reason to believe that if Maxwell's equations are true in a given frame, then pure electromagnetic measuring sticks are shrunk when moving relative to this frame and pure electromagnetic clocks are slowed. Electromagnetism is so constructed that pure electromagnetic matter does not allow you to detect a preferential frame at all, and provided measurements are made with electromagnetically constructed instruments, Maxwell's equations will be equally valid in all inertial frames, but coordinates so measured will not be related by the Galilean transformations.

Natural reactions:

(1) Horribly complicated.

It is indeed, but this is the way the Lorentz transformations were first obtained. It is fascinating to read the accounts of these heroic struggles given by Lorentz and by Larmor — despite the (charmingly) archaic phraseology, they speak to the heart of anyone who has lived through the last twenty years of particle physics.

(2) But matter is not pure electromagnetic anyway.

True, but perhaps nature has organised a conspiracy so that we can never detect these effects with anything? Then we could forget about preferred frames. We make measurements with real rods and clocks and in motion they do funny things relative to Galilean space. Maxwell's equations (interpreted in terms of real measurements with apparatus constructed from electromagnetism) are covariant under a set of transformations other than the Galilean — the Lorentz transformations. If there is a universal conspiracy, we may adopt the point of view that velocity does the same funny things to all clocks and rods or we may work directly in terms of real measurements and relate the coordinates of an event measured in two different inertial frames through the Lorentz transformations rather than the Galilean. These two points of view are operationally indistinguishable. We shall investigate the second — special relativity. We shall find the Lorentz transformations of course, and a corollary. If a principle of relativity embodying the Lorentz transformations holds universally, then Newtonian mechanics requires modification.

GENERAL REFERENCES

Lorentz's viewpoint may be found in a paper (1904) reprinted in

The Principle of Relativity, A. Einstein et al (Dover 1952) — still in print.

It is also discussed in

The Theory of Electrons, H. A. Lorentz (2nd ed. Leipzig, New York 1916)

Larmor's work is contained in

Aether and Matter, J. Larmor (Cambridge 1900)

These two books are relatively rare but can be found in some libraries. A modern account of this approach is given by

J. S. Bell, *Prog. Scientific Culture* **1/2** 135 (1976)

Lecture 2 THE LORENTZ TRANSFORMATIONS

In the first lecture we more or less defined the concept of a principle of relativity, defined inertial frames, investigated the Galilean transformations and Newton's laws and then passed on to electromagnetism. We found that even in an aether theory there are good grounds for expecting the null result of the Michelson-Morley experiment because the theory predicts the contraction of measuring rods (and the slowing of clocks) travelling through the aether. The contraction also fixes up the force between moving strings of charge and in no case is motion with respect to the aether observable. If all forms of matter are identically affected by velocity, then we may retain a principle of relativity.

We retain the idea of inertial frames, in which particles which are not acted on by any force move with constant velocity and in which physics does not depend on either the origin or orientation of the coordinate system. We postulate that Maxwell's equations take the same form, with the same numerical content, in all such inertial frames, and seek a relation between measured coordinates (x_i, t) in one inertial frame S and the measured coordinates of the same event (x'_i, t') in another frame S', such that this condition is satisfied; in particular the measured velocity of light is a universal constant. In special relativity we further postulate that all inertial frames are equivalent, that the whole of physics is governed by equations equally true for all unaccelerated observers and that these equations retain their form and numerical content under the transformations we are seeking. Because we have learned that in an aether theory funny things happen to rods and clocks as a function of velocity, the transformations we seek need not accord with our intuition (but they had better reduce to the Galilean transformations in the low velocity limit).

Philosophically, we are shifting our view of space and time from one in which they are marked off in absolute units, to which a key exists, to an operational view in which space is that which is measured with measuring sticks and time is that which is measured with clocks. It is convenient and appropriate, seems obvious once spelt out, but is not strictly necessary.

(x_i, t) are the coordinates of an event measured by a given observer and his assistants in an inertial frame S, using standard instruments. (For example, the unit of length could be 10^8 ^{12}C atoms, the unit of time a reciprocal atomic frequency, and clocks at different places in S synchronised using the laws of physics obtaining in S). The same event (make it a supernova explosion if you want to be spectacular) has coordinates (x'_i, t') measured by an observer in S', using instruments constructed according to the same physical specification as those in S. What must the relation be between (x_i, t) and (x'_i, t') if both observers find Maxwell's equations are true?

There is no correct way of deriving these transformations (the Lorentz transformations). Lorentz discovered them by playing about with Maxwell's equations, a long and hard route. The important thing is the output, not the input. It is perfectly in order to make plausible guesses, so long as the answer is satisfactory. A rigorous derivation can only proceed from a set of assumptions, usually with the benefit of hindsight, and here we shall work from two very simple assumptions. The first (abstracted from Maxwell's equations) is that the speed of light is a universal constant c, measured in any inertial frame. The second assumption is that the transformations are linear in (x_i, t), (x'_i, t').

A general transformation is

$$x'_i = f_i(x_j, t)$$
$$t' = f_4(x_j, t)$$

A linear transformation is

$$\begin{aligned} x' &= a_{11}x + a_{12}y + a_{13}z + a_{14}t \\ y' &= a_{21}x + a_{22}y + a_{23}z + a_{24}t \\ z' &= a_{31}x + a_{32}y + a_{33}z + a_{34}t \\ t' &= a_{41}x + a_{42}y + a_{43}z + a_{44}t \end{aligned} \tag{2.1}$$

where the coefficients $a_{\mu\nu}$ do not depend on $(x_i,\, t)$. This assumption ensures that a particle moving with constant velocity in an inertial frame S is also measured as having (different) constant velocity in a different inertial frame S'.

[PROBLEM: Prove that the linearity assumption is sufficient to ensure this. Is it strictly necessary?]

We may write eq.(2.1) in shorthand:

$$x'_\mu = a_{\mu\nu}x_\nu \tag{2.2}$$

where the greek indices run from 1 to 4 and we adopt the convention of summation over repeated indices. Conversely we can write

$$x_\nu = (a^{-1})_{\mu\nu}x'_\mu \qquad \text{where} \qquad (a^{-1})_{\mu\nu}a_{\mu\nu} = 1$$

where we can explicity SOLVE (2.1) for x_ν in terms of the quantities x'_μ in order to determine the inverse a^{-1}.

The quantities $a_{\mu\nu}$ will depend on the relative velocity of the two frames (and in general on their orientation) and we can enormously restrict the possibilities by a limited use of our first assumption.

For events on an electromagnetic wavefront, we have the relations

$$\left(\frac{d\mathbf{x}\prime}{dt'}\right)\cdot\left(\frac{d\mathbf{x}'}{dt'}\right) = c^2 = \left(\frac{d\mathbf{x}}{dt}\right)\cdot\left(\frac{d\mathbf{x}}{dt}\right) \tag{2.3}$$

Then because of rotational invariance (which we wish to keep) we MUST have

$$\begin{aligned} \mathbf{x}' &= \mathbf{x} + \mathbf{V} \\ t' &= t + S \end{aligned}$$

where $\mathbf{V}$ is some vector and S is some scalar quantity. However, $\mathbf{V}$ and S must be linear in $\mathbf{x}$, t, but may be functions of the relative velocity $\mathbf{v}$. The available vectors are

$$\mathbf{x}, \mathbf{v}t, \mathbf{v}(\mathbf{v}.\mathbf{x})$$

and the available scalars are

$$t, \mathbf{v}.\mathbf{x}$$

For convenience restrict the axes so that x, x' are common and y, y'; z, z' coincide at $x = x' = 0$, $t = t' = 0$ and x is along $\mathbf{v}$. (There is no restriction on the physics implied by this choice so long as the origin and orientation of axes have no physical significance). Then the components of the three available vectors are

$$(x, y, z), (vt, 0, 0), (v^2x, 0, 0)$$

and the scalars are t and vx. (Any of these quantities may be multiplied by scalar functions of the relative velocity).

Therefore we see at once that

$$y', z' \qquad \text{cannot depend on} \qquad x, t$$

$$x', t' \qquad \text{cannot depend on} \qquad y, z$$

for our choice of parallel axes (we can always later rotate either or both sets of axes to reach other configurations).

Remember that v is the velocity with which a fixed point in S' moves along the positive x axis in S:

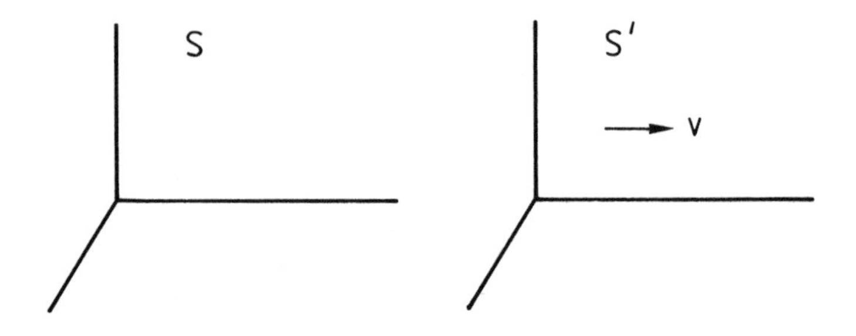

The observer in S sees a fixed point in S' receding with measured velocity v, while an observer in S' sees a fixed point in S receding along the (negative) x' axis with velocity v, if relative velocity has any meaning.

For parallel axes, we have already restricted the possible transformations to the form

$$\begin{aligned} x' &= a_{11}x + a_{14}t \\ y' &= a_{22}y, \quad z' = a_{33}z \\ t' &= a_{41}x + a_{44}t \end{aligned} \tag{2.4}$$

The relations (2.4) could quite properly be taken as plausible starting assumptions, but from the rotational invariance of (2.3) we already know that

$$a_{14}t \quad \propto \quad vt$$

$$a_{41}x \quad \propto \quad vx$$

$$a_{11}, a_{44}, a_{22}, a_{33} \quad \text{may be scalar functions of } v.$$

Since a_{22}, a_{33} are only scalar functions of velocity, and may not contain (x_i, t), their inverse must be a_{22}, a_{33}. Solving for y, z in terms of y', z' we have

$$y = y'/a_{22}, \qquad z = z'/a_{33}$$

so

$$a_{22} = \pm 1, \qquad a_{33} = \pm 1$$

and we selected the positive values through our original choice of axes.

We now have the relations

$$\begin{aligned} x' &= a_{11}x + a_{14}t \\ y' &= y, \quad z' = z \\ t' &= a_{41}x + a_{44}t \end{aligned} \tag{2.5}$$

(which is also a plausible set of starting assumptions).

Now

$$\left.\frac{dx}{dt}\right|_{S'} = v \quad ; \quad \left.\frac{dx'}{dt'}\right|_{S} = -v$$

Taking appropriate differentials of (2.5) we have

$$\left.\frac{dx'}{dt}\right|_{S'} = 0 = a_{11}\left.\frac{dx}{dt}\right|_{S'} + a_{14} = a_{11}v + a_{14}$$

Then

$$a_{14} = -va_{11} \tag{2.6}$$

(which is also true for the Galilean transformations)

$$\left.\frac{dx'}{dt'}\right|_{S} = -v = a_{14}\left.\frac{dt}{dt'}\right|_{S} \tag{2.7}$$

$$\begin{aligned} \left.\frac{dt'}{dt}\right|_{S} &= a_{44} = -\frac{a_{14}}{v} \qquad \text{(from (2.7))} \\ &= a_{11} \qquad \text{(from (2.6))} \end{aligned}$$

Thus using the definition of relative velocity we have the transformations

$$\begin{aligned} x' &= a_{11}(x - vt) \\ y' &= y; z' = z \\ t' &= a_{41}x + a_{11}t \end{aligned} \tag{2.8}$$

With $a_{11} = 1$, $a_{41} = 0$, these relations are the Galilean transformations. We now inject the universal value of the velocity of light.

In each frame, a spherical wave spreading from the origin has coordinates (x_i, t), (x'_i, t') such that

$$\begin{aligned} x^2 + y^2 + z^2 - c^2t^2 &= 0 \\ x'^2 + y'^2 + z'^2 - c^2t'^2 &= 0 \end{aligned}$$

For these special events on an expanding wavefront,

$$\begin{aligned} x^2 - c^2t^2 &= -(y^2 + z^2) \qquad |x| \le ct \\ x'^2 - c^2t'^2 &= -(y'^2 + z'^2) \qquad |x'| \le ct' \end{aligned}$$

but we already have the general relations

$$y' = y; \quad z' = z$$

For a range of x, t and a range of values of $x^2 - c^2t^2$ we have

$$x^2 - c^2t^2 = x'^2 - c^2t'^2$$

and given that the transformations are linear, we must have

$$x^2 - c^2t^2 \equiv x'^2 - c^2t'^2$$

regardless of the value of $x^2 - c^2t^2$. The identity CANNOT be established from the relations

$$x^2 - c^2t^2 = 0$$
$$x'^2 - c^2t'^2 = 0$$

for propagation along the x axes.

Given the identity,

$$x^2 - c^2t^2 \equiv a_{11}^2(x - vt)^2 - c^2(a_{41}x + a_{11}t)^2$$

we may equate coefficients of x^2, t^2, xt to obtain

$$\begin{aligned} 1 &= a_{11}^2 - c^2a_{41}^2 \\ 0 &= -a_{11}^2v - c^2a_{41}a_{11} \\ -c^2 &= v^2a_{11}^2 - c^2a_{11}^2 \end{aligned}$$

Three equations, two unknowns. Fortunately they are consistent, with

$$\begin{aligned} a_{11} &= \frac{1}{\sqrt{1 - v^2/c^2}} \\ a_{41} &= \frac{-1}{\sqrt{1 - v^2/c^2}}\frac{v}{c^2} \end{aligned}$$

and we finally have

$$\begin{aligned} x' &= \frac{x - vt}{\sqrt{1 - v^2/c^2}} \\ y' &= y; z' = z \\ t' &= \frac{t - vx/c^2}{\sqrt{1 - v^2/c^2}} \end{aligned} \tag{2.9}$$

These are the Lorentz transformations for the particular case of parallel axes and coincident origins.

[PROBLEM: Obtain the more general relations for parallel axes but an arbitrary direction of relative motion

$$\mathbf{x}' = \mathbf{x} - \frac{\mathbf{v}(\mathbf{v}.\mathbf{x})}{v^2}\left(1 - \frac{1}{\sqrt{1 - v^2/c^2}}\right) - \frac{\mathbf{v}t}{\sqrt{1 - v^2/c^2}}$$
$$t' = \frac{t - \mathbf{v}.\mathbf{x}/c^2}{\sqrt{1 - v^2/c^2}}$$

(In practice it is usual to choose coordinate systems with parallel axes, one axis coincident with the relative velocity vector).]

The defining property of the Lorentz transformations is that the quantity

$$S_{12}^2 = (x_1 - x_2)^2 + (y_1 - y_2)^2 + (z_1 - z_2)^2 - c^2(t_1 - t_2)^2$$

is an *invariant*, having the SAME value in all inertial frames, where the x, y, z, t intervals are measured with real instruments constructed according to the same specifications in every frame. The invariant interval S_{12} is zero only for light signals, but is always invariant. Note that different observers will measure different space and time intervals individually between the same two events.

Take the Lorentz transformations (2.9) giving x', y', z', t' in terms of x, y, z, t and solve for x, y, z, t in terms of x', y', z', t', obtaining

$$\begin{aligned} x &= \frac{x' + vt'}{\sqrt{1 - v^2/c^2}} \\ y &= y'; z = z' \\ t &= \frac{t' + vx'/c^2}{\sqrt{1 - v^2/c^2}} \end{aligned} \tag{2.10}$$

which is the result which would be obtained by swapping the primes in (2.9) and reversing the sign of $\mathbf{v}$. This seems sensible. Equations (2.9) and (2.10) in fact have the same form and numerical content, because

$$\left.\frac{dx}{dt}\right|_{S'} = v \quad ; \quad \left.\frac{dx'}{dt'}\right|_{S} = -v$$

and the transformations can be written

$$x' = \left(x - \left.\frac{dx}{dt}\right|_{S'} t\right) \Big/ \sqrt{1 - \left(\left.\frac{dx}{dt}\right|_{S'}\right)^2 /c^2} \qquad x = \left(x' - \left.\frac{dx'}{dt'}\right|_{S} t'\right) \Big/ \sqrt{1 - \left(\left.\frac{dx'}{dt'}\right|_{S}\right)^2 /c^2}$$

$$y' = y; z' = z \qquad y = y'; z = z'$$

$$t' = \left(t - \left.\frac{dx}{dt}\right|_{S'} \frac{x}{c^2}\right) \Big/ \sqrt{1 - \left(\left.\frac{dx}{dt}\right|_{S'}\right)^2 /c^2} \qquad t = \left(t' - \left.\frac{dx'}{dt'}\right|_{S} \frac{x'}{c^2}\right) \Big/ \sqrt{1 - \left(\left.\frac{dx'}{dt'}\right|_{S}\right)^2 /c^2}$$

which differ only in swapping the primes—an arbitrary labelling. Thus the transformations are reciprocal and are themselves covariant, having the same form and numerical content in all inertial frames.

If the hypothesis of universal special relativity is correct, then these transformations will work equally well whatever physical processes are used to define units of space and time—for example lengths measured in terms of the diameter of an atomic nucleus and time measured with clocks driven by β-radioactivity. We shall proceed to look at the observable consequences, starting with the phenomena of time dilation and Lorentz contraction.

GENERAL REFERENCES

Anyone intrigued by the game of deriving the Lorentz transformations from minimal assumptions should consult

The Theory of Space, Time and Gravitation. V. Fock (Pergamon, 1966)

Principles of Relativity Physics. J.L. Anderson (Academic Press, 1967)
(Section entitled 'Derivation of the Poincaré mappings')

Lecture 3 TIME DILATION AND LORENTZ CONTRACTION

In the second lecture we derived from a rather minimal set of assumptions the Lorentz transformations for parallel axes, relative motion along the x axis, and common origins:

$$
\begin{aligned}
x' &= \frac{x - vt}{\sqrt{1 - v^2/c^2}} & x &= \frac{x' + vt'}{\sqrt{1 - v^2/c^2}} \\
y' &= y; \quad z' = z & y &= y'; \quad z = z' \\
t' &= \frac{t - vx/c^2}{\sqrt{1 - v^2/c^2}} & t &= \frac{t' + vx'/c^2}{\sqrt{1 - v^2/c^2}} \\
v &= \left.\frac{dx}{dt}\right|_{x'} & v &= -\left.\frac{dx'}{dt'}\right|_{x}
\end{aligned}
\tag{3.1}
$$

Remember that (x, y, z, t) are the coordinates (of a given event) relative to the origin, measured with real apparatus in S, (x', y', z', t') are the coordinates of the SAME event as measured in S'. Note that the transformations do not make sense for a measured relative velocity exceeding c. We obtained these relations by ensuring that a particle with constant (measured) velocity in one frame has constant (measured) velocity in the other, and further demanding that the (measured) velocity of light is a universal constant.

The characteristic distinguishing the Lorentz transformations is that the interval between two events is an invariant:

$$\Delta x^2 + \Delta y^2 + \Delta z^2 - c^2\Delta t^2 \equiv \Delta x'^2 + \Delta y'^2 + \Delta z'^2 - c^2\Delta t'^2 \tag{3.2}$$

[PROBLEM: Show this explicitly from (3.1)]

This invariant can be positive (spacelike), zero (light cone) or negative (timelike).

If the interval is *Spacelike* you can always find a velocity $v < c$ such that $\Delta t' = 0$ (starting in S with given $\Delta \mathbf{x}$, Δt) but you can never find a velocity $v < c$ such that $\Delta \mathbf{x}' = 0$.

[PROBLEM: Prove these assertions]

If the interval is *Timelike* you can always find a velocity $v < c$ such that $\Delta \mathbf{x}' = 0$, but you can never make $\Delta t' = 0$.

[PROBLEM: Prove these assertions]

Thus two observers in different inertial frames will only agree that two events are simultaneous if they occur at the same physical location in space (coalescing into a single event).

Note that two events with a spacelike separation cannot be linked by signals travelling at the speed of light, the implication being that two such events cannot causally affect one another. (If the sun blows up NOW, you have another 8 minutes left). Events with a timelike separation can be linked by light signals and so can be causally connected. For timelike separations, however, if Δt is positive, $\Delta t'$ in any other inertial frame is also positive, while if Δt is negative, $\Delta t'$ is always negative. Thus any point in spacetime has an absolute past and an absolute future.

In two dimensions, x and t, we can draw a space-time diagram:

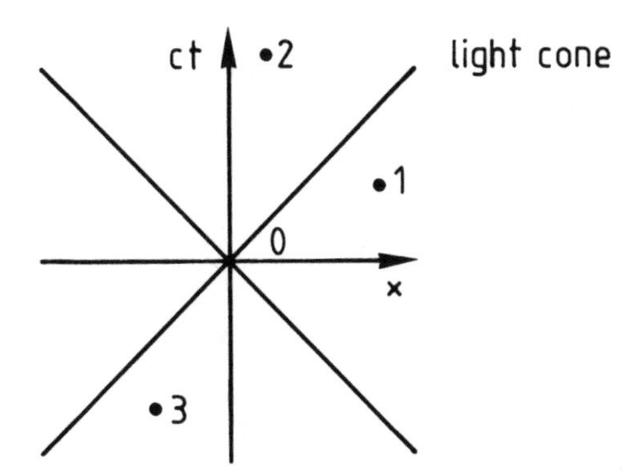

Events separated by a spacelike interval from 0 cannot affect or be affected by event 0 at all, for example event 1. Lorentz transformations move the relative position of event 1 around but always keep it outside the light cone. Events separated by a timelike interval from 0 can be causally connected with 0: Event 2 lies in the absolute future of 0 and Event 3 in the absolute past. Lorentz transformations move events such as 2 and 3 around, but only within their respective segments of the light cone.

[PROBLEM: For given (x_1, t_1), (x_2, t_2), (x_3, t_3) in S, find curves on which these events lie in terms of (x', t') by eliminating (arbitrary) $v < c$]

With this preamble, we proceed to one of the apparently most curious results of the hypotheses of special relativity, *Time dilation*.

Consider two events from a sequence which in a particular frame occur at the same place, such as ticks of a single clock at rest in a frame S'. Then their spatial separation in S' is $\Delta\mathbf{x}' = 0$. This frame is the *rest frame* of this sequence of events and the time interval in S' between two such events is the *proper time* interval, denoted by $\Delta\tau$. Then from the invariant interval (2)

$$c^2\Delta\tau^2 = c^2\Delta t^2 - \Delta x^2 - \Delta y^2 - \Delta z^2 \tag{3.3}$$

Divide by Δt^2 (where Δt is the time interval between the two events measured in some other frame S) and obtain

$$\left(\frac{\Delta\tau}{\Delta t}\right)^2 = 1 - \frac{v^2}{c^2} \tag{3.4}$$

because $\frac{d\mathbf{x}}{dt}\big|_{S'} = \mathbf{v}$ where $\mathbf{v}$ is the velocity of a fixed point in S' measured in S.

Therefore

$$\Delta\tau = \Delta t\sqrt{1 - v^2/c^2} \tag{3.5}$$

The proper time interval between two events is always less than the time interval measured, between the same two events, in any other frame. The longer time interval is the one measured with two clocks at different places, synchronised in their own rest frame. It should be at once clear from (3.2) and (3.3) that had we chosen a sequence of events with $\Delta\mathbf{x} = 0$, then $\Delta t'$ would have been greater than Δt.

In fact we have already encountered this result in the last lecture, when we found

$$\left.\frac{dt'}{dt}\right|_S = a_{44} = \frac{1}{\sqrt{1 - v^2/c^2}}$$

(where Δt is an element of proper time because the relation specifies a fixed position in S).

These results can also be obtained directly from the Lorentz transformations, of course. For example

$$\Delta t' = \frac{\Delta t - \Delta x v/c^2}{\sqrt{1 - v^2/c^2}} \tag{3.6}$$

If Δx is chosen to be zero, then Δt is an element of proper time and $\Delta t'$ is greater. If however you are considering a situation in which the two events occur at the same place in S', then there are two ways to solve the problem. First, you may note that for a fixed position in S', $\Delta x/\Delta t = v$ and substitute into (3.6). Secondly you may take the inverse transformation

$$\Delta t = \frac{\Delta t' + \Delta x' v/c^2}{\sqrt{1 - v^2/c^2}} \tag{3.7}$$

and set $\Delta x' = 0$. Either way the same result is obtained. Time dilation is symmetric between the two frames, and is defined quite unambiguously: the shortest measured time interval between two events obtains in their rest frame. Such a frame can always be found for two events with a timelike separation. In practical applications, the rest frame usually has an obvious significance.

EXAMPLE: Consider an individual unstable particle which in its own rest frame lives a time τ_1. In the laboratory frame it has velocity v and its measured lifetime in the laboratory will be $\tau_1/\sqrt{1 - v^2/c^2}$.

The lifetime of an individual particle cannot be predicted in any frame, but the distribution of lifetimes in any one frame is exponential. Consider a single particle, say a positive K meson (kaon), produced when the primary beam of a proton synchroton strikes a target. It passes through a counter at x_1 in the laboratory frame and at x_2 it decays in a detector. The time interval in the laboratory between these two events is measured by comparing the arrival time of a decay signal from the detector with the arrival time of a signal from the first counter, the signals being brought to some central station through known lengths of time calibrated cable. We are effectively measuring the time interval with two clocks at different places in the laboratory, synchronised in the laboratory. The time interval in the laboratory is

$$\Delta t = \frac{x_2 - x_1}{v}$$

(where v is the velocity of the particle in the laboratory) and the *Proper time* interval is

$$\Delta \tau = \sqrt{1 - v^2/c^2}\,\Delta t$$

If we sample a large number of K^+, with the same velocity, then in the rest frame the distribution of lifetimes is given by

$$\frac{dN}{d\tau} = \frac{N_0}{\tau_M} e^{-\tau/\tau_M}$$

where τ_M is the proper mean lifetime (often simply called the proper lifetime). In the laboratory, then, substituting

$$\tau = \sqrt{1 - v^2/c^2}\,t, \qquad d\tau = \sqrt{1 - v^2/c^2}\,dt$$

$$\frac{dN}{dt} = N_0 \frac{\sqrt{1 - v^2/c^2}}{\tau_M} e^{-\sqrt{1 - v^2/c^2}\,t/\tau_M}$$

or

$$\frac{dN}{dt} = \frac{N_0}{t_M} e^{-t/t_M}$$

where t_M is the mean lifetime measured in the laboratory and is related to the proper liftime by

$$\tau_M = \sqrt{1 - v^2/c^2}\, t_M$$

We may evaluate the probability of a given K^+ surviving from the counter at x_1 to a given point x_2; it is

$$e^{-\frac{x_2 - x_1}{v t_M}}$$

It doesn't matter whether the kaons come through in a bunch, or scattered in time; the results apply provided the beam has been velocity selected (in practice, momentum selected). If the velocity varies too, then the time (or distance) distribution in the laboratory will not be exponential, but provided the lifetime and velocity of each particle is measured, the rest frame lifetime can be calculated and the distribution of decay times in the rest frame is exponential.

Time dilation is verified every day in high energy accelerator laboratories: secondary beams of pions and kaons are routinely produced with the factor $1/\sqrt{1 - v^2/c^2}$ varying from ~ 1 up to values ≥ 1000. Without any special effort, experiments studying secondary particles or utilising secondary beams check time dilation at the few per cent level. The phenomenon makes accessible for study decay lifetimes and decay processes which otherwise could hardly be detected.

[PROBLEM: Beams of pions and kaons can be readily produced with energies ~ 200 GeV. The proper mass of the charged pion is 0.14 GeV/c^2, proper lifetime $2.6 \times 10^{-8} s$ and the proper mass of the charged kaon is 0.494 GeV/c^2, proper lifetime $1.2 \times 10^{-8} s$. Calculate the laboratory lifetimes of these particles when they have energy 200 GeV (in these units, the ratio of proper mass to energy is the factor $\sqrt{1 - v^2/c^2}$). How far will 200 GeV pions and kaons travel before each species is attenuated by a factor e due to decay?]

[PROBLEM: The heavy lepton τ (the third copy of the electron) has a mass of 1.784 GeV/c^2. $\tau^+\tau^-$ pairs are produced with colliding beams of equal energy in the process $e^+e^- \rightarrow \tau^+\tau^-$. What is the distribution of the distance between the τ^+ and τ^- decay points if the beam energy is 20 GeV and the proper τ lifetime is $3 \times 10^{-13} s$? (Answer: Exponential, with a mean of 0.2 cm.)]

It may seem strange that elapsed time is different in two frames and in particular that an observer O measures O′'s clocks to be running slower than his (constructed according to the SAME recipe) and VICE VERSA. Strange it may be, but there is no element of paradox. A clock at a given place in one frame is being compared with two clocks, synchronised in their own rest frame but at different places in that rest frame. In frame S the time difference between two clocks is recorded, in S' the time interval is measured by one clock, and VICE VERSA.

We will return to this problem at the end of the lecture, but now move on to another phenomenon which seems curious because of its reciprocal character, *Lorentz Contraction*.

If you want to measure the length of a rod which is at rest in the laboratory, you may unroll a tape measure alongside it and record the readings at the two ends. They do not change as a function of time. If a rod is moving through the laboratory, you have to think a bit more carefully about what is meant by a measurement of its length. For low velocities there is no problem—you simply take a photograph with a flash of light and because the velocity is low you don't worry about the travel time of the light. Effectively you are recording the positions of the two ends at the same instant of laboratory time and the length is the difference between the two coordinates. This is what is meant by the length of a moving rod. If you were to record the position of one end at $t = 0$ and of the other at $t = 5$s, a correction would be made for the velocity of the rod; the distance travelled by the rod in 5s would be subtracted.

Return to eqs.(3.1), giving the relation between measured coordinates of an event in two frames, with a common origin for the coordinates. For two events, we may write

$$\Delta x' = \frac{\Delta x - v\Delta t}{\sqrt{1 - v^2/c^2}}$$

Let the two events be the measurements of the two ends of a rod, at rest in S', made in S simultaneously. Then the time interval Δt is zero (although the time interval $\Delta t'$ is not). The separation between these two events in S' is the proper length of the rod, which can be measured and remeasured (in S') walking from end to end with a tape measure. The spatial separation between these two events in S is

$$\Delta x = \Delta x' \sqrt{1 - v^2/c^2} < \Delta x' \tag{3.8}$$

and this is the length of the rod measured in S, in which it is moving. A moving rod is measured as being shortened in the direction of motion. This is the Lorentz contraction, in the framework of special relativity now rather than in Lorentz's electron theory. It is immediately obvious from (3.1) that the relation is reciprocal—if we choose $\Delta t'$ to be zero instead of Δt, then $\Delta x' = \sqrt{1 - v^2/c^2}\Delta x$. Again there is no paradoxical element—the frame in which the rod is measured as contracted is selected by demanding that the two measurements are made simultaneously in the frame in which it is moving. The Lorentz contraction is not susceptible to direct measurement (at least at present) in the same way as time dilation, but it is a physical effect as real as time dilation and necessarily linked to it. It must be remembered that (3.8) relates the results of two carefully specified measurements. The Lorentz contraction contributes to the visual appearance of a rapidly moving (macroscopic) object, but is not the whole story, for the eye (or camera) records light arriving simultaneously at the retina (or film) from many points and for rapidly moving objects the travel time and relative direction of light rays in the source and observer frames are important.

[PROBLEM: If you have a taste for such things, show that a cube travelling at high velocity normal to one of its faces looks rotated through an angle $\sim v/c$].

To illustrate the relation between time dilation and Lorentz contraction, it is enough to note that

$$\left.\frac{dx}{dt}\right|_{S'} = v \quad ; \quad \left.\frac{dx'}{dt'}\right|_{S} = -v$$

As a concrete example, consider once more an unstable particle which passes through a counter at $x = 0$, $t = 0$ in the laboratory and decays at x, t in a second counter. In its rest frame, the particle remains at $x' = 0$ and lives a time t'. Then using eqs.(3.1)

$$x' = \frac{x - vt}{\sqrt{1 - v^2/c^2}} = 0 \qquad \text{Hence } x = vt \text{ of course}$$

$$t' = \frac{t - vx/c^2}{\sqrt{1 - v^2/c^2}} = t\sqrt{1 - v^2/c^2} \qquad \text{(time dilation again, using the formula the other way round)}$$

In time t' the position of the first counter has moved back a distance vt' in the particle rest frame. This is the distance between the two counters measured in the particle rest frame. But

$$vt' = vt\sqrt{1 - c^2/c^2} = x\sqrt{1 - v^2/c^2}$$

where x is the distance surveyed between the two counters on the floor of the experimental hall.

Apparent inconsistencies abound, but they are all fallacious. Consider for example the following situation, again involving two counters on the floor of an experimental hall. Things are so arranged that at $t = t' = 0$, $x = x' = 0$ two kaons are produced in the first counter, one moving at high velocity and the other at rest. Suppose each lives a proper time τ, the one produced in motion with respect to the first counter decaying in the second counter. Then applying time dilation, the kaon which decays in the second counter lives a time τ in its own rest frame and a time $\tau/\sqrt{1 - v^2/c^2}$ in the laboratory. The kaon which is created and decays in the first counter lives a time τ in the laboratory and a time $\tau/\sqrt{1 - v^2/c^2}$ in the frame of the meson moving with respect to the lab. Thus the time ordering of the two decays is reversed between the two relevant frames:

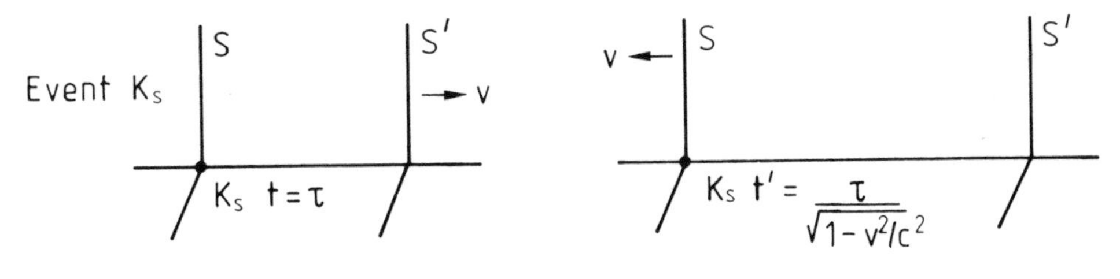

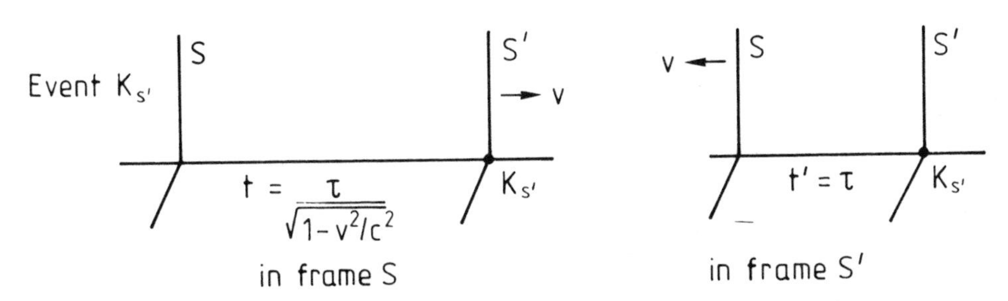

It seems inconsistent—but it isn't provided the two events depicted in the figure—K_S and $K_{S'}$—are spacelike separated, for remember that for spacelike separated events the time ordering may be different in different frames. Spacelike separated events cannot influence each other through any signal propagated with velocity $\leq c$, and provided no signals can propagate with velocity $> c$ no logical paradoxes arise. If signals could propagate faster than light (the ultrawaves of science fiction) then clock readings for clocks constructed from ordinary matter could no longer be regarded as measuring time in the same way and we would be forced back in the direction of Lorentz's electron theory.

[PROBLEM: Show explicitly that the two events K_S, $K_{S'}$ are spacelike separated]

[PROBLEM: There is a hoary old favourite involving Lorentz contraction which is posed as follows: A ruler of proper length 1m slides over a table in which a slot of proper length

1m is cut. The length of the ruler is < 1m in the table frame; the length of the slot is < 1m in the ruler frame. Does the ruler fall through? As posed, the problem contains a lot of irrelevant garbage—the ruler will start to tilt under gravity as its centre of mass clears the edge of the slot. So clean it up by having the table and the ruler in empty space, both very thin, and finally introduce a tiny component of velocity perpendicular to the table. Don't worry about the speed of pressure waves in the ruler if the leading edge strikes—the question is simply whether or not the ruler can go through the slot without either end striking the table. (The answer is that it can go through if its proper length is less than the proper length of the slot and it can't if its proper length is greater than the proper length of the slot)]

Lecture 4 INVARIANTS, 4-VECTORS AND COVARIANCE

This lecture is concerned with formal mathematical techniques. They are very easy to pick up and are almost essential for a proper grasp of the physics of special relativity. In particular the modifications needed to Newtonian mechanics are obvious once these techniques have been mastered and understood within their physical setting. We assume from now on that the Lorentz transformations between measured coordinates in two different inertial frames are correct, and that we have a universal principle of relativity.

We first define *invariants*. A physical quantity which has the same value for all inertial observers in *invariant* under the Lorentz transformations, and called an *invariant*, or *4-scalar*. Examples are: the invariant interval between two events (proper time is a special case), the phase of a wave, proper (or rest) mass, and electric charge.

Coordinates (or relative coordinates) x, y, z, t are made up from an ordinary vector x, y, z and a timelike piece t. The quantity

$$x^2 + y^2 + z^2 - c^2t^2$$

is invariant under the Lorentz transformations, just as

$$x^2 + y^2 + z^2$$

is invariant under rotations and translations. If we define coordinates

$$x_\mu = \mathbf{x}, ict \qquad (\mu \text{ running from 1-4})$$

then the defining property of the Lorentz transformations is

$$\sum_\mu x_\mu x_\mu = \text{invariant}$$

We now introduce the convenient convention of summation over repeated indices

$$x_\mu x_\mu \equiv \sum_\mu x_\mu x_\mu$$

and express the Lorentz transformations in the compact form

$$x'_\mu = a_{\mu\nu} x_\nu \tag{4.1}$$

The invariance of the interval between two events with relative coordinates x_μ then requires that

$$\underset{\text{sum over } \mu}{x'_\mu x'_\mu} = \underset{\text{sum over } \mu,\rho,\sigma}{a_{\mu\rho} x_\rho a_{\mu\sigma} x_\sigma} \equiv \underset{\text{sum over } \nu}{x_\nu x_\nu} \tag{4.2}$$

The $a_{\mu\nu}$ depend only on the relative velocity of the two frames of reference and are essentially contained in eqs.(3.1) for the special case of parallel axes and coincident origins, although in general they also contain rotations and translations.

From (4.2) we can see that the $a_{\mu\nu}$ must satisfy the relation

$$a_{\mu\rho} a_{\mu\sigma} = \delta_{\rho\sigma}$$

where $\delta_{\mu\nu}$ is the Krönecker δ function

$$\begin{array}{llll} \delta_{\mu\nu} = 1 & \text{for} & \mu = \nu & \text{(any value)} \\ \delta_{\mu\nu} = 0 & \text{for} & \mu \neq \nu & \end{array}$$

(This is easy to see: $\underset{\text{sum over } \mu,\rho,\sigma}{a_{\mu\rho}x_\rho a_{\mu\sigma}x_\sigma} = \underset{\text{sum over } \rho,\sigma}{\delta_{\rho\sigma}x_\rho x_\sigma} = \underset{\text{sum over } \nu}{x_\nu x_\nu}$)

Any physical quantity which has three components forming a vector and a fourth timelike component, the four components being mixed together under the Lorentz transformations according to

$$V'_\mu = a_{\mu\nu}V_\nu$$

is called a *4-vector*. Its square is automatically invariant:

$$V'_\mu V'_\mu = a_{\mu\rho}V_\rho a_{\mu\sigma}V_\sigma = \delta_{\rho\sigma}V_\rho V_\sigma = V_\nu V_\nu$$

[NOTE: A lot of different ways of representing 4-vectors and their squares are in use. Very often the timelike component is x_0 (μ running from 0-3) rather than x_4 and very often the square is defined as $c^2t^2 - \mathbf{x}^2$. The use of time as an imaginary coordinate is becoming unpopular: for example one may define two quantities

$$x_\mu = (ct, \mathbf{x})$$
$$x^\mu = (ct, -\mathbf{x})$$

and take for the square of a 4-vector $x^\mu x_\mu$, or one may define the square of a vector to be

$$g_{\mu\nu}x_\mu x_\nu$$

where

$$g_{\mu\nu} = \begin{pmatrix} 1 & & & \\ & -1 & & \\ & & -1 & \\ & & & -1 \end{pmatrix}$$

When working in the framework we have established—flat space, linearly calibrated instruments—none of this matters in the slightest, so long as you know what you are doing, which is, when squaring a 4-vector, to sum the squares of the spacelike components and subtract from that sum the square of the timelike component (with an overall positive or negative sign added according to taste—you may like to have the square of a timelike invariant interval positive). If you wish to employ general curvilinear coordinates then the introduction of $g_{\mu\nu}$ is mandatory, and in the relativistic theory of gravitation (general relativity) the gravitational potentials end up in a $g_{\mu\nu}$].

Let's now extract some physics by applying these techniques. We first note that the 4-scalar product of two different 4-vectors is also an invariant:

$$A'_\mu B'_\mu = a_{\mu\rho}A_\rho a_{\mu\sigma}B_\sigma = A_\nu B_\nu$$

Now the phase of a wave is

$$\mathbf{k}\,.\,\mathbf{x} - \omega t = \mathbf{k}\,.\,\mathbf{x} + \left(i\frac{\omega}{c}\right)(ict)$$

where ω is the angular frequency and **k** is the propagation vector. The phase is an invariant, $(\mathbf{x}, ict)$ is a 4-vector and therefore

$$k_\mu = \left(\mathbf{k}, i\frac{\omega}{c}\right)$$

must also be a 4-vector. We can therefore trivially write down the transformation properties of the wave vector and frequency, by a comparison with eqs.(3.1):

$$\begin{aligned} k'_x &= \frac{k_x - v\omega/c^2}{\sqrt{1 - v^2/c^2}} \\ k'_y &= k_y; \qquad k'_z = k_z \\ w' &= \frac{\omega - vk_x}{\sqrt{1 - v^2/c^2}} \end{aligned} \tag{4.3}$$

For an electromagnetic wave, $k = \frac{\omega}{c}$ and so $k_x = \frac{\omega}{c}\cos\theta$ where θ is the angle in S between the propagation vector and the x axis, defined as being the axis of relative motion of S' to S. Then

$$\omega' = \frac{\omega(1 - v/c\cos\theta)}{\sqrt{1 - v^2/c^2}} \tag{4.4}$$

which is the relativistic Doppler effect. Note that there is a second order effect even for the transverse case ($\theta = 90\,\text{deg}$) which of course reflects time dilation. This leads us to a practical problem: which way round should (4.4) be used? The answer is simple. The frequency ω and the angle θ are measured in the same frame. Thus if you are measuring light from a star and the light is propagating at right angles to the direction of motion of the star, then ω is measured and ω' is the proper frequency—the frequency of the spectral line in the rest frame of the star under observation. The frequency you measure on earth is reduced (the line is red shifted) by just the amount predicted by time dilation.

[PROBLEM: Light from a star is observed at 90 deg to the direction of motion of a star and the measured frequency is ω. Calculate the proper frequency of this spectral line (trivial) AND the direction in which the observed light was emitted in the rest frame of the star (Answer: $\cos^{-1}(-{}^{v}/_{c})$). Now suppose that light of proper frequency ω is emitted by the star at 90 deg to the line of motion in the rest frame of the star. What is the observed frequency and what angle does the observed light make with the relative direction of motion of the star and the earth?]

You could of course have obtained the relativistic Doppler shift, eq.(4.4), by requiring the phase to be an invariant and applying the treatment given in the first lecture, using Lorentz transformations rather than Galilean.

Note that the invariant

$$k_\mu k_\mu = k^2 - \omega^2/c^2$$

is zero for electromagnetic radiation $\omega/k = \omega'/k' = c$ when k_μ is a *null* 4-vector.

[PROBLEM: (This one is both real and interesting: see B. Margon, Scientific American Oct. 1980 p44)

The bizarre spectrum of the object SS433 (about 12,000 light years from earth) exhibits very strong H_α emission lines with large Doppler shifts. Both red and blue shifts are present simultaneously and the red and blue shifts vary, in phase. The extreme red shift is to a wavelength of 7680Å, coincident in time with the extreme blue shift to a wavelength of 5913Å. After just over 50 days, the lines merge at a common wavelength of 6797Å. Subsidiary extremes are reached after about 30 more days, at wavelengths of 7102Åand 6490Å. The proper wavelength of the H_α line is 6563Å.

A simple model which works very well is to suppose that some compact central object (probably a neutron star or black hole accreting gas from a companion) is emitting two opposite narrow jets of gas at high velocity. The overall period of 164 days presumably represents a precession of the jet axis. Discuss this model in the light of the data given and find the velocity of the gas in the jets (Answer: 0.26c), together with any other significant parameters you can infer].

In addition to 4-scalars and 4-vectors we can define 4-scalar and 4-vector fields.

Let $\phi(x_\alpha)$ be a 4-scalar field, which means that ϕ varies from point to point but at a given space-time point has the same value for all inertial observers. Then, just as in the algebra of vector fields $\nabla\phi$ is a vector field, the four quantities

$$\frac{\partial \phi}{\partial x_\mu}$$

are the four components of a 4-vector field, and if $V_\mu(x_\alpha)$ is a 4-vector field,

$$\frac{\partial V_\mu}{\partial x_\mu}$$

is a scalar field—a 4-divergence.

This looks obvious but we had better prove it. For a given relative velocity between two frames S and S', x'_μ is a function only the four coordinates x_ν so we can write

$$\frac{\partial}{\partial x'_\mu} = \frac{\partial x_\nu}{\partial x'_\mu}\frac{\partial}{\partial x_\nu}$$

(remember the summation over all values of ν).

Now

$$x'_\mu = a_{\mu\nu} x_\nu \tag{4.1}$$

and we want to find $\frac{\partial x_\nu}{\partial x'_\mu}$. If we differentiate (1) we are left with a sum of these terms over all μ: we need to invert (1). So multiply (1) by $a_{\mu\rho}$ and carry out the summation over all μ:

$$a_{\mu\rho} x'_\mu = a_{\mu\rho} a_{\mu\nu} x_\nu = \delta_{\rho\nu} x_\nu = x_\rho$$

So

$$a_{\mu\nu} x'_\mu = x_\nu \tag{4.5}$$

and differentiating with respect to one particular component x'_μ we find

$$a_{\mu\nu} = \frac{\partial x_\nu}{\partial x'_\mu} \tag{4.6}$$

Thus

$$\frac{\partial}{\partial x'_\mu} = a_{\mu\nu}\frac{\partial}{\partial x_\nu}$$

and is a 4-vector operator.

Then

$$\frac{\partial \phi'}{\partial x'_\mu} = \frac{\partial \phi}{\partial x'_\mu} = \frac{\partial \phi}{\partial x_\nu}\frac{\partial x_\nu}{\partial x'_\mu} = a_{\mu\nu}\frac{\partial \phi}{\partial x_\nu}$$

The four derivatives with respect to the spacetime coordinates of a scalar field are mixed up under a Lorentz transformation in exactly the same way as the components of a 4-vector:

$$\frac{\partial \phi}{\partial x_\mu} \quad \text{is a 4-vector field.}$$

Similarly,

$$\frac{\partial V'_\mu}{\partial x'_\mu} = \frac{\partial}{\partial x'_\mu}(a_{\mu\rho}V_\rho) = \frac{\partial x_\sigma}{\partial x'_\mu}a_{\mu\rho}\frac{\partial V_\rho}{\partial x_\sigma}$$
$$= a_{\mu\sigma}a_{\mu\rho}\frac{\partial V_\rho}{\partial x_\sigma} = \delta_{\sigma\rho}\frac{\partial V_\rho}{\partial x_\sigma} = \frac{\partial V_\nu}{\partial x_\nu}$$

and $\frac{\partial V_\mu}{\partial x_\mu}$ is indeed a 4-scalar field.

The operator

$$\Box = \frac{\partial^2}{\partial x_\mu \partial x_\mu}$$

is a scalar operator. If ϕ is a scalar field, then $\Box\phi$ is a scalar field, while if V_μ is a vector field, $\Box V_\mu$ is a vector field.

[PROBLEM: Prove these assertions]

This is known as the d'Alembertian and is a generalisation of the Laplacian ∇^2: $\Box = \nabla^2 - \frac{1}{c^2}\frac{\partial^2}{\partial t^2}$

So far, this is formal mathematics, but we shall now put it to work. Remember that we have merely found that the velocity of light is invariant under the Lorentz transformations, and we have not shown that Maxwell's equations keep the same form in all inertial frames.

A convenient place to start is with the equation of continuity:

$$\nabla \cdot \mathbf{J} + \frac{\partial \rho}{\partial t} = 0$$

where $\mathbf{J}$ is current density and ρ charge density. This expresses local conservation of charge—integrating over volume and applying Gauss' theorem we obtain

$$\frac{\partial Q}{\partial t} = -\int_S \mathbf{J} \cdot \mathbf{dS}$$

where Q is the charge contained within the closed surface S. Charge only disappears from some volume by flowing outward through the bounding surface. Note that if charge is conserved and special relativity is valid, the conservation law must be local. Charge cannot disappear here and simultaneously reappear a mile away, because space-like separated events are not simultaneous in all inertial frames.

Now $\mathbf{J}$ is an ordinary vector field, so we are tempted to define a new 4-vector field

$$J_\mu = (\mathbf{J}, ic\rho)$$

and write the equation of continuity in the form

$$\frac{\partial J_\mu}{\partial x_\mu} = 0$$

If J_μ is indeed a 4-vector field, then this equation is not only covariant but manifestly so. But J_μ contains real physics, not just mathematical abstractions. If J_μ is a 4-vector field, then the charge on a particle (as opposed to the charge density) must be an invariant.

Suppose a particle is at rest in a frame S', in which it has charge density ρ'. (Let the particle be a proton if you are worried by the point-like nature of the electron). Then the charge on this particle is given by

$$q' = \int \rho' dx' dy' dz'$$

If $ic\rho$ is the fourth component of a 4-vector, then

$$\rho' = \frac{(\rho - \frac{\mathbf{v.J}}{c^2})}{\sqrt{1 - v^2/c^2}}$$

where in the frame S the charge and current densities are ρ and $\mathbf{J}$, and the particle is moving with velocity $\mathbf{v}$ in S. Now $\mathbf{J} = \rho\mathbf{v}$, so we find at once that

$$\rho' = \rho\sqrt{1 - v^2/c^2}$$

(and you could get this also by writing

$$\rho = \frac{\left(\rho' + \frac{\mathbf{v.J'}}{c^2}\right)}{\sqrt{1 - v^2/c^2}}$$

and setting $\mathbf{J}' = 0$ for S' to be the rest frame of the particle).

At a given instant of time the charge of the particle in S is

$$q = \int \rho dx dy dz = \int \frac{\rho'}{\sqrt{1 - v^2/c^2}} dx dy dz$$

Now we may equate spatial intervals transverse to the motion

$$dy' = dy; \qquad dz' = dz$$

and since we are picking a given instant of time in S the distance dx corresponding to dx' in S' is Lorentz contracted

$$dx = \sqrt{1 - v^2/c^2}\, dx'$$

As a result

$$q = \int \rho dx dy dz = \int \rho' dx' dy' dz' = q'$$

If J_μ is a 4-vector field, we have a covariant law of (local) conservation of charge, AND charge is an invariant. Experimentally, charge IS an invariant: the force on a particle moving in a pure electric field is independent of the velocity.

[PROBLEM: Think of observations which test this].

Now examine the electromagnetic field equations. It is convenient first to write them using potentials

$$\begin{aligned} \nabla^2 \mathbf{A} - \frac{1}{c^2}\frac{\partial^2 \mathbf{A}}{\partial t^2} &= -\frac{4\pi}{c}\mathbf{J} \\ \nabla^2 \phi - \frac{1}{c^2}\frac{\partial^2 \phi}{\partial t^2} &= -4\pi\rho \end{aligned} \tag{4.7}$$

where this form depends on choosing $\mathbf{A}$ such that

$$\nabla \cdot \mathbf{A} + \frac{1}{c}\frac{\partial \phi}{\partial t} = 0 \tag{4.8}$$

(choosing the Lorentz gauge).

Eq. (4.8) looks like the 4-divergence of a 4-vector, while (4.7) can be written as

$$\begin{aligned} \Box \mathbf{A} &= -\frac{4\pi}{c}\mathbf{J} \\ \Box \phi &= -4\pi\rho \end{aligned}$$

If we define a new 4-vector field

$$A_\mu = (A, i\phi)$$

Then (4.7) and (4.8) become

$$\Box A_\mu = -\frac{4\pi}{c}J_\mu \quad ; \quad \frac{\partial A_\mu}{\partial x_\mu} = 0$$

which are manifestly covariant. In particular, if we transform

$$\Box' A'_\mu = -\frac{4\pi}{c}J'_\mu$$

we obtain

$$\Box a_{\mu\nu} A_\nu = -\frac{4\pi}{c} a_{\mu\sigma} J_\sigma$$

which is shorthand for a set of 4 equations (one for each value of μ) which can be solved to yield

$$\Box A_\mu = -\frac{4\pi}{c} J_\mu$$

—same form, same numerical content in both frames.

The whole trick of writing covariant equations is to ensure that all components have the same transformation properties: 4-scalar=4-scalar, 4-vector=4-vector .... Such equations retain their form and numerical content under the Lorentz transformations, just as equations of the forms scalar=scalar, vector=vector ... retain their form and numerical content under rotations in ordinary 3-space.

We know that J_μ is a 4-vector field. If (4.7), (4.8) are to be covariant under the Lorentz transformations, then the potentials A, ϕ must transform as components of a 4-vector. This can of course be checked. Work in a frame where Maxwell's equations are known to hold and solve (4.7) for a particle at rest and for a particle (with the same charge) in uniform motion. Then check that the potentials for the charge in motion are identical to taking the potentials for the charge at rest and transforming under the assumption that $(A, i\phi)$ is a 4-vector. It works.

What about the electric and magnetic fields? The potentials are defined so as to obtain

$$\mathbf{E} = -\nabla\phi - \frac{1}{c}\frac{\partial \mathbf{A}}{\partial t}$$
$$\mathbf{B} = \nabla \times \mathbf{A}$$

so

$$\begin{aligned} E_j &= i\left(\frac{\partial A_4}{\partial x_j} - \frac{\partial A_j}{\partial x_4}\right) \\ B_i &= \frac{\partial A_k}{\partial x_j} - \frac{\partial A_j}{\partial x_k} \end{aligned} \tag{4.9}$$

The components of $\mathbf{E}$ and $\mathbf{B}$ are thus the six independent components of the second rank antisymmetric tensor

$$F_{\mu\nu} = \frac{\partial A_\nu}{\partial x_\mu} - \frac{\partial A_\mu}{\partial x_\nu}$$

which under the Lorentz transformations clearly transforms as

$$F'_{\mu\nu} = a_{\mu\rho}a_{\nu\sigma}F_{\rho\sigma}$$

The original Maxwell equations can then be written in manifestly covariant form in terms of the field tensor $F_{\mu\nu}$.

We have not so far attempted a Lorentz covariant description of the equation of motion of a charged particle in an electromagnetic field. We shall come to that after investigating the modifications which we have to make to particle kinematics: momentum, energy and of course $E = mc^2$.

Lecture 5 MOMENTUM, ENERGY, KINEMATICS AND DYNAMICS

The equations of electromagnetism are covariant with respect to the Lorentz transformations and are not covariant with respect to the Galilean transformations. However, the laws of Newtonian mechanics are covariant with respect to the Galilean transformations. Experimentally the equations of electromagnetism hold in all inertial frames and if a principle of relativity covers the whole of physics, the same transformations must work equally well in all parts of physics and these transformations must be the Lorentz transformations.

This is not ruled out by the successes of Newtonian mechanics, for these successes applied in a regime $v \ll c$, and in this regime the Lorentz transformations reduce to the Galilean transformations.

We embark, therefore, on a search for new laws of mechanics which reduce in the low velocity limit to Newton's laws, but which are covariant with respect to the Lorentz transformations. In the end of course these laws must be compared with experiment in order to check whether a principle of relativity really holds. A clear indication that Newtonian concepts are in need of modification is provided by the fact that the momentum of a charged particle increases faster than linearly with velocity in a frame in which Maxwell's equations hold, due to the compression of the electromagnetic field surrounding it.

The keystone of mechanics is conservation of momentum and energy, and we suppose that there exist physical quantities which we may identify with momentum and energy which are conserved in all inertial frames. We need to find such quantities: they must reduce to the familiar low velocity quantities and admit of writing conservation of energy and momentum in a manifestly covariant way.

We start by introducing a new 4-vector. The coordinates of a particle, viewed from a given inertial frame, constitute the prototype 4-vector $x_\mu = (x, y, z, ict)$. Let τ be the proper time associated with this particle (the time measured by a standard clock moving with the particle). Then we can define a new 4-vector, the 4-velocity

$$V_\mu = \frac{dx_\mu}{d\tau}$$

This is a 4-vector because dx_μ is the (infinitesimal) difference of two 4-vectors and $d\tau$ is an invariant. It is important to note that while the space components of the 4-velocity reduce to ordinary velocity in the low velocity limit, $d\tau \to dt$, they are NOT identical with ordinary velocity. Rather, we have

$$\begin{aligned} \mathbf{v} &= \frac{d\mathbf{x}}{dt} = \frac{d\mathbf{x}}{d\tau}\frac{d\tau}{dt} \qquad \text{so } \mathbf{V} = \frac{\mathbf{v}}{\sqrt{1 - v^2/c^2}} \\ V_4 &= \frac{dx_4}{d\tau} = ic\frac{dt}{d\tau} = \frac{ic}{\sqrt{1 - v^2/c^2}} \end{aligned}$$

where

$$V_\mu = (\mathbf{V}, V_4)$$

[PROBLEM: Check these relations by starting from the rest frame of the particle, where $V_\mu = (0, 0, 0, ic)$ and transforming to a frame in which the particle moves with (ordinary) velocity $\mathbf{v}$. (What is the value of the invariant $V_\mu V_\mu$?)]

The utility of the 4-velocity lies in its being a 4-vector: $V'_\mu = a_{\mu\nu} V_\nu$. Ordinary velocity is a ratio of components of a 4-vector and has more complicated transformation properties, which may be easily obtained from the Lorentz transformations, eqs. (3.1). If x'_μ, x_μ are the coordinates of a particle at a given point in space-time, then since

$$x' = \frac{x - vt}{\sqrt{1 - v^2/c^2}} \qquad y' = y, \qquad z' = z \qquad t' = \frac{t - v\,x/c^2}{\sqrt{1 - v^2/c^2}}$$

then

$$\frac{dx'}{dt'} = \frac{dx - v\,dt}{dt - v\,dx/c^2} \quad , \quad \frac{dy'}{dt'} = \frac{dy\sqrt{1 - v^2/c^2}}{dt - v\,dx/c^2}$$

or

$$u'_x = \frac{u_x - v}{1 - \frac{v\,u_x}{c^2}} \quad , \quad u'_y = \frac{u_y\sqrt{1 - v^2/c^2}}{1 - \frac{v\,u_x}{c^2}} \tag{5.1}$$

where it should be remembered that we have chosen parallel axes with motion along the common x axis.

[PROBLEM: If you are still bothered about the recipricocity of the transformations of special relativity, solve (5.1) for u_x and u_y in terms of u'_x, u'_y]

[PROBLEM: Let a particle have 4-velocity U'_μ in S'. Find the velocity $\mathbf{u}'$. Then transform U'_μ to the frame S, eliminate the parameter τ and hence obtain (5.1)]

[PROBLEM: show that the velocity transformations (5.1) leave the speed of light invariant.]

[PROBLEM: The relativistic addition of velocities embodied in (5.1) seems peculiar when phrased in familiar terms: A fly is travelling with velocity u' relative to the still air in the cabin of an airliner travelling with velocity v in the same direction, yet the velocity of the fly relative to the ground is $< u' + v$ (and if we replace the fly by a pulse from a laser, the velocity is c in both frames). Qualitatively the physical reason is obvious: u' is measured aboard the airliner by timing the fly over a distance $\Delta x'$, taking a time $\Delta t'$. The measured distance between the two markers is Lorentz contracted, and the time $\Delta t'$ is dilated. We might expect

$$\Delta x = v\Delta t + \Delta x'\sqrt{1 - v^2/c^2} \tag{i}$$

and

$$\Delta t = \Delta t'/\sqrt{1 - v^2/c^2} \tag{ii}$$

and hence $u < u' + v$.

However, while (i) is right, (ii) is not - check this by extracting an expression for u from (i) and (ii) and comparing it with the result extracted from (5.1). Find the flaw in the above argument and correct it.]

We may similarly define a 4-acceleration $d^2x_\mu/d\tau^2$ and use these new 4-vectors to write relativistically covariant equations of motion. For the present, the importance of the 4-velocity lies in its application to the problem of relativistic energy and momentum. We note that momentum is a vector with three components and that energy is related to momentum....

At low velocity, $dt \to d\tau$ and $\mathbf{V} \to \mathbf{v}$, ordinary velocity. We formally define a new 4-vector by multiplying the 4-velocity by a quantity with the dimensions of mass which is an invariant. The new 4-vector, called for obvious reasons the 4-momentum, is

$$p_\mu = m_0\frac{dx_\mu}{d\tau}$$

and as the velocity goes to zero the spacelike components reduce to the familiar non-relativistic momentum. The quantity m_0 is thus the inertial mass exhibited by our test particle in its own rest frame, the rest mass, or proper mass, which is inherently an invariant quantity.

The 4-momentum of a set of i particles is

$$P_\mu = \sum_i p^i_\mu = \sum_i m^i_o \frac{dx^i_\mu}{d\tau^i}$$

and is also a 4-vector, by construction. Note that the element of proper time, $d\tau^i$, is specific to each particle i.

We now suppose that a set of i particles undergo collisions in some local region, and a set of j particles emerge with different 4-momenta. (The outgoing particles need not be identical with the incoming particles, and the number need not be the same). The total change in 4-momentum, making all measurements in a single inertial frame, is

$$\Delta P_\mu = \sum_j m^j_o \frac{dx^j_\mu}{d\tau^j}(\text{out}) - \sum_i m^i_o \frac{dx^i_\mu}{d\tau^i}(\text{in})$$

and is again a 4-vector.

If the 4-momentum is conserved in any one inertial frame S,

$$\Delta P_\mu = 0 \tag{5.2}$$

(remember, this is shorthand for a set of 4 equations, setting each of the four components equal to zero).

In some other inertial frame, S',

$$\Delta P'_\mu = a_{\mu\nu} \Delta P_\nu$$

Thus if the 4-momentum is conserved in any one inertial frame, it is conserved in all inertial frames, because each component of $\Delta P'_\mu$ is made up of a sum of the four components of ΔP_μ, each of which is zero. Eq. (5.2) is covariant. If p_μ contains momentum and energy, then we have written the property of conservation of momentum and energy in a manifestly covariant form.

Now

$$\begin{aligned} p_\mu &= m_o \frac{dx_\mu}{d\tau} = m_o \frac{dx_\mu}{dt}\frac{dt}{d\tau} \\ \mathbf{p} &= m_o \frac{d\mathbf{x}}{dt}\frac{1}{\sqrt{1-v^2/c^2}} = \frac{m_o \mathbf{v}}{\sqrt{1-v^2/c^2}} \\ p_4 &= m_o \frac{dx_4}{dt} = icm_o \frac{dt}{d\tau} = \frac{icm_o}{\sqrt{1-v^2/c^2}} \end{aligned} \tag{5.3}$$

If we define

$$p_4 = \frac{iE}{c}$$

then

$$E = \frac{m_o c^2}{\sqrt{1-v^2/c^2}} \tag{5.4}$$

and E has the dimensions of energy. The 4-momentum of a particle is then

$$p_\mu = (\mathbf{p}, iE/c)$$

Expand the quantities $\mathbf{p}$, E for small v/c:

$$\mathbf{p} = m_0\mathbf{v} + O\left(\frac{v}{c}\right)^3 \ldots$$
$$E = m_0c^2 + \frac{1}{2}m_0v^2 + O\left(\frac{v}{c}\right)^4 c^2 \ldots$$

The second term in E is the non-relativistic kinetic energy, T. In low energy kinematics, where the mass m_0 does not change perceptibly in collisions, conservation of E implies conservation of T, and conservation of $\mathbf{p}$ implies conservation of non-relativistic momentum. We therefore identify the quantities $\mathbf{p}$, E with (relativistic) energy and momentum and complete relativistic kinematics by postulating that these quantities are conserved. The conservation laws are Lorentz covariant. The quantities $\mathbf{p}$, E appearing in the 4-momentum p_μ satisfy the relation

$$p_\mu p_\mu = -m_0^2c^2 \qquad \text{i.e.} \qquad p^2 - E^2/c^2 = -m_0^2c^2$$

Because the 4-momentum p_μ is a 4-vector,

$$p'_\mu = a_{\mu\nu}p_\nu$$

and the relations between momentum and energy as measured in two frames S' and S, with relative velocity v along the x axis, take the form (compare with (3.1), (4.3))

$$p'_x = \frac{p_x - vE/c}{\sqrt{1 - v^2/c^2}}$$
$$p'_y = p_y \qquad p'_z = p_z$$
$$E' = \frac{E - vp_x}{\sqrt{1 - v^2/c^2}}$$

[PROBLEM: Solve these equations for $\mathbf{p}$, E in terms of $\mathbf{p}'$, E' and check that the result is sensible.]

If we DEFINE momentum as the product of velocity and inertial mass

$$\mathbf{p} = m\mathbf{v}$$

then

$$m = \frac{m_0}{\sqrt{1 - v^2/c^2}} = E/c^2 \tag{5.5}$$

The rest energy of particle is thus $E_0 = m_0c^2$ and the total energy is $E = mc^2$ where m is the inertial mass. The inertial mass of a particle rises with velocity and while mass corresponds to energy, energy has inertia. This is the mechanism which ensures that a particle cannot be accelerated to a velocity greater than that of light. There are thus two aspects to the famous relation $E = mc^2$. On the one hand, the inertial mass of a particle increases with its energy. On the other, an energy $E_0 = m_0c^2$ resides in a particle at rest (the rest mass energy). Our covariant conservation laws guarantee only that the sum of rest mass energy and kinetic energy is a conserved quantity, which suggests that energy may extracted by reducing the sum of rest masses of a system.

[PROBLEM: The energy flux reaching the earth from the sun is 2 cal cm^{-2} min^{-1}. How much mass does the sun lose each second? Do you expect any effect on planetary orbits? (The mass of the sun is 2×10^{33}gm)]

[PROBLEM: The mass of the hydrogen atom is 1.6735×10^{-24}gm. The mass of the helium atom is 6.6456×10^{-24}gm. The sun is a main sequence star, powered by the conversion of hydrogen to helium. How many tonnes of helium are produced each second within the sun?]

Note that special relativity makes no predictions about the feasibility of reducing the mass of a system and getting energy (let alone useful energy) out. It is necessary to understand the structure and dynamics of any specified system before such a prediction can be made. However, given that the inertial mass of a nucleus (measured by mass spectroscopy) is less than the summed mass of its constituents, special relativity implies that energy will be released on formation of that nucleus from its constituents.

Given that the mass of the uranium nucleus is greater than the sum of the masses of plausible fission fragments, then energy will be released if a uranium nucleus can be persuaded to undergo fission. Furthermore, in this instance there is some understanding of the mechanism involved. It is the repulsive coulomb forces that are responsible for the greater relative mass of the uranium nucleus and consequently fission fragments will be driven apart by the coulomb field and acquire as kinetic energy most of the energy represented by the mass difference. Such energy is useful, but relativity alone does not imply that it is possible to tap the stored energy of the uranium nucleus.

[PROBLEM: The useful energy released in the fission of ^{235}U is $\sim$ 200 MeV. How many uranium nuclei undergo fission each second in a 1000 MW reactor complex?]

The mass-energy equivalence (in the second sense) was first checked out in the early days of nuclear physics, the masses of nuclei being determined through mass spectroscopy and the energy released in reactions involving nuclei being directly measured. [The relativistic expressions for momentum and energy, together with the relation between inertial mass and energy, are essentially contained within Maxwell's equations for the specific case of electromagnetism, for both the energy and momentum stored in the electromagnetic fields surrounding a charged particle, either moving or at rest, can be calculated. The momentum flux in a plane electromagnetic wave is equal to the energy flux divided by c: the photon has rest mass zero. For an abstract pure electromagnetic matter, the relativistic expressions for momentum, energy and inertial mass would obtain in any frame in which Maxwell's equations hold.]

We have so far defined energy and momentum in kinematic terms and we have not considered particle dynamics. The balance between incoming and outgoing energy and momentum in the interaction of high energy particles is a problem in kinematics—the acceleration of charged particles to reach high energy is a dynamical problem.

We can see that we expect the equations of motion of a particle to take the covariant form

$$F_\mu = m_0 \frac{d^2 x_\mu}{d\tau^2} = \frac{dp_\mu}{d\tau} \tag{5.6}$$

where F_μ is a 4-vector, the 4-force, and $d^2x_\mu/d\tau^2$ is the 4-acceleration. The Lorentz force acting on a charged particle takes precisely this form. The particle is now accelerating and we have introduced the 4-acceleration, where the coordinates x_μ of the particle are always measured in one particular inertial frame (any inertial frame you like, but you don't change the inertial frame of (5.6) as a function of time). The infinitesimal quantity

$d\tau$ is an infinitesimal element of proper time measured in the instantaneous rest frame of the particle. This frame DOES change with time.

In most practical applications, electric and magnetic fields are employed to accelerate charged particles and all that is wanted is the equation of motion in the laboratory frame. Maxwell's equation's plus the Lorentz force provide almost everything, but the proper application of the definition of momentum and energy is essential. In Newtonian mechanics we have two possible definitions of force which are equivalent: mass $\times$ acceleration and rate of change of momentum. How should we write relativistic equations of motion? The answer is, of course, as in eq. (5.6) but we would wish to dispense with the parameter τ.

Consider a particle accelerated by a force which is constant in the rest frame of the particle. In the instantaneous rest frame, the fourth component of the 4-acceleration must be zero, because in that frame $dt = d\tau$. The three space-like components of the 4-acceleration are, in the instantaneous rest frame, the three components of ordinary acceleration (which may be non-zero even though the velocity in the rest frame is always zero).

[PROBLEM: What is the scalar product of the 4-velocity with the 4-acceleration?]

If we have a 4-force

$$f_\mu = (f_1, 0, 0, 0)$$

in the instantaneous rest frame, then in some other inertial frame

$$F_1 = \frac{f_1}{\sqrt{1 - v^2/c^2}} \quad ; \quad F_4 = \frac{i\,v\,f_1/c}{\sqrt{1 - v^2/c^2}}$$

so $F_4 = i\,\mathbf{v}.\mathbf{F}/c$

(F_1 and F_4 are not independent, $F_\mu F_\mu = f_\mu f_\mu$)

In the observer's frame, we would therefore write

$$\frac{dp_1}{d\tau} = F_1 = \frac{f_1}{\sqrt{1 - v^2/c^2}}$$

$$\frac{dE}{d\tau} = \mathbf{v}.\mathbf{F} = \frac{v f_1}{\sqrt{1 - v^2/c^2}}$$

The proper time τ is not directly measured. Setting

$$d\tau = dt\sqrt{1 - v^2/c^2}$$

where v is the instantaneous velocity of the particle at the space-time point where it has momentum $\mathbf{p}$, energy E and experiences a force f_1 in its own rest frame, we have

$$\frac{dp_1}{dt} = f_1 \quad ; \quad \frac{dE}{dt} = f_1 v$$

and more generally

$$\frac{d\mathbf{p}}{dt} = \mathbf{f} \quad ; \quad \frac{dE}{dt} = \mathbf{f}.\mathbf{v} \tag{5.7}$$

where $\mathbf{f}$ is the proper force, and is identical to the space-like 4-force in the instantaneous rest frame. Eq. (5.7) is not manifestly covariant, but it follows from eq. (5.6) which is. It should be noted that while $d\mathbf{p}/d\tau$, $dE/d\tau$ make up a 4-vector, $d\mathbf{p}/dt$ and dE/dt do not. The first three components of the 4-force give the rate of change of momentum with proper time and the fourth component gives the rate of change of energy with proper time. The relation between the two parts of (5.7) is not unexpected, for the work done by a force $\mathbf{f}$ moving a distance $d\mathbf{x}$ is $\mathbf{f}.d\mathbf{x}$, so the rate of change of energy is $\mathbf{v}.\mathbf{f}$ (but see the note at the end of this lecture).

We can obtain the acceleration easily:

$$\begin{aligned}\frac{d\mathbf{p}}{dt} &= \frac{d}{dt}(m\mathbf{v}) = \frac{d}{dt}\left(\frac{E}{c^2}\mathbf{v}\right)\\ &= \frac{E}{c^2}\frac{d\mathbf{v}}{dt} + \frac{\mathbf{v}}{c^2}\frac{dE}{dt}\end{aligned}$$

Using (5.7), (5.3) and (5.4)

$$m_0\frac{d\mathbf{v}}{dt} = \left(\mathbf{f} - \frac{\mathbf{v}(\mathbf{v}.\mathbf{f})}{c^2}\right)\sqrt{1 - v^2/c^2} \tag{5.8}$$

If the dependence of $\mathbf{f}$ on t is known, then either (5.7) or (5.8) may be integrated to find the final velocity.

[PROBLEM: For a constant force f, integrate (5.8) from zero velocity at time $t = 0$ and find the maximum velocity achieved.]

Thus relativistically the ordinary force is to be equated to the rate of change of momentum of a particle and is not to be equated to the product of the acceleration and either the proper mass or the inertial mass.

[PROBLEM: Instead of working from eq. (5.7) we could have set

$$\begin{aligned}m_0\frac{d^2x}{d\tau^2} &= \frac{f_1}{\sqrt{1 - v^2/c^2}}\\ m_0\frac{d^2t}{d\tau^2} &= \frac{v/c^2 f_1}{\sqrt{1 - v^2/c^2}}\end{aligned}$$

and obtained the acceleration (5.8) by eliminating τ directly. Do it, just to convince yourself that there was no hanky-panky.]

The equation of motion of a charged particle in a pure electric field $\mathbf{E}$ is thus to be written

$$\frac{d\mathbf{p}}{dt} = e\,\mathbf{E}$$

where e is the charge of the particle (remember e is an invariant).

[PROBLEM: The SLAC linear accelerator produces electrons of 20 GeV (4×10^4 rest mass energy)

(i) Calculate their velocity at exit
(ii) If the electrons experience a constant force eE and are accelerated over 3km in the laboratory, find the field strength E
(iii) How far would electrons have to travel in this field in order to reach the same velocity if Newtonian mechanics applied? (Answer: 3.8 cm)
(iv) Estimate the length of the accelerator as measured in the electron rest frame.

For a measurement of the electron velocity, see Z.G.T. Guiragossian et al, Phys. Rev. Lett. **34** *335* (1975)]

We can now easily check the covariance of the equations of motion in an electromagnetic field. The Lorentz force law is

$$\mathbf{F} = e\big(\mathbf{E} + \frac{1}{c}\mathbf{v} \times \mathbf{B}\big)$$

The electric and magnetic fields are obtained from the 4-curl of the 4-vector potential A_μ, and we expect

$$\frac{d\mathbf{p}}{dt} = e\big(\mathbf{E} + \frac{1}{c}\mathbf{v} \times \mathbf{B}\big) \quad ; \quad \frac{dE}{dt} = e\mathbf{E}.\mathbf{v} \tag{5.9}$$

The curl of the potential is a second rank tensor and must be contracted with a 4-vector to yield a 4-force. The appropriate 4-vector is clearly the 4-velocity, and the expression

$$\frac{dp_\nu}{d\tau} = \frac{e}{c}\frac{dx_\mu}{d\tau}\left(\frac{\partial A_\mu}{\partial x_\nu} - \frac{\partial A_\nu}{\partial x_\mu}\right)$$

is both manifestly covariant and reduces to (5.9) on making the replacements $\frac{d}{dt} \to \frac{dt}{d\tau}\frac{d}{dt}$; $\mathbf{B} = \nabla \times \mathbf{A}, \mathbf{E} = -\big(\nabla\phi + \frac{1}{c}\frac{\partial \mathbf{A}}{\partial t}\big)$

[PROBLEM: For circular motion in the laboratory, an acceleration v^2/r must be applied at right angles to the direction of the velocity $\mathbf{v}$. Prove that the condition for a relativistic particle to execute circular motion of radius r in a (constant) magnetic field B is

$$p = \frac{e}{c}Br$$

where p is the magnitude of the momentum of a particle of charge e.

In the project originally known as the Desertron, currently under discussion in the U.S.A., it is proposed to collide beams of protons circulating in opposite directions at energies of up to 10 TeV (10^4 GeV). Calculate the radius of the machine assuming

(i) iron-cored magnets, maximum field $\sim$ 15kG (1.5 tesla)
(ii) Super-conducting magnets, maximum field $\sim$ 60kG (6 tesla) (Superconducting magnets are favoured and the machine is now known as the Super-conducting Super Collider, or SSC)]

Quite apart from direct verification of mass energy equivalence with atomic nuclei to an accuracy of about one part in 10^6, the relations of relativistic kinematics are verified daily in the experiments of high energy physics. Particle accelerators are designed in accord with the prescriptions of special relativity, and work. For example, the Stanford linear accelerator operates with a travelling electromagnetic wave in a waveguide loaded so that the wave speeds up to keep pace with the surf-riding electrons, and the electron velocity is calculated from (5.9), (5.3) and (5.4). The energy acquired by the electrons can of course be measured, if desired, with a calorimeter. At present electron beams exist with energies in excess of 20 GeV ($E/m_ec^2 > 4 \times 10^4$) at the SLAC and DESY laboratories, and the highest energy beam of protons will shortly be the 1000 GeV beam from the Tevatron at Fermilab ($E/m_pc^2 > 10^3$). Relativistic kinematics is simply a tool of the high energy physicist's trade.

[NOTE: The interpretation of covariant equations of motion is not in general as straightforward as it is in the case of electromagnetism. This point is usually neglected in elementary treatments of special relativity, because high energy particles are invariably handled with electromagnetic fields, described by a 4-vector potential, the charge being an invariant. If Lorentz covariant equations of motion are written for motion in 4-scalar or 4-tensor fields, complications set in, and the rest mass is potential dependent. For example, a covariant equation of motion for a particle in a scalar potential would be of the form

$$\frac{dp_\mu}{d\tau} = -g\frac{\partial \phi}{\partial x_\mu}$$

if g is an invariant charge. Explore the consequences of this equation for the case where the potential ϕ is, in a particular frame, time independent.

So far as is known, there are no long range scalar fields. Electromagnetism and gravitation are both long range, and the gravitational potential is a 4-tensor. The relativistic theory of gravitation can be written in a purely Lorentz covariant way, but the complications are considerable. In a Lorentz covariant theory of gravitation, charged particles and electromagnetic fields, mass is a function of the local gravitational potentials, light travels at a speed determined by the local gravitational potentials and hence neither travels in straight lines nor at constant speed, and both the rate at which clocks run and the proper length of standard measuring sticks are affected by the gravitational potentials ... but none of these effects is locally observable (see M.G. Bowler, Gravitation and Relativity *Pergamon* 1976)]

GENERAL REFERENCES

Less abstract arguments leading to the relativistic expressions for momentum, energy and inertial mass may be found in, for example

The Feynman Lectures on Physics **Vol. I, Ch. 16** (*Addison-Wesley* 1963)
J. D. Jackson, Classical Electrodynamics **Ch. 11** (*2nd Ed., Wiley* 1975)

The discussions given by Einstein in his early papers are founded in electromagnetism—see A. Einstein et. al., The Principle of Relativity (*Dover*).

Lecture 6 TRICKS OF THE TRADE

The applications of special relativity, uncontaminated by the complications of the gravitational field, lie in areas where terms $\sim \frac{v^2}{c^2}$ are significant and yet the gravitational potential may be regarded as constant. To a large extent this means particle physics. (Flying atomic clocks around the world is sensitive to the change of gravitational potential as well as to velocity). Theoretically, it is axiomatic that any new theory should be Lorentz covariant, while the experimentalist is primarily concerned with the manipulations of relativistic kinematics.

In the last lecture we constructed the 4-momentum

$$p_\mu = m_0 \frac{dx_\mu}{d\tau} \quad ; \quad p_\mu = \left(\mathbf{p}, i\frac{E}{c}\right)$$

and noted that the spacelike components reduce to the familiar Newtonian momentum in the low velocity limit, while the quantity E reduces to the Newtonian kinetic energy, plus an additional term m_0c^2, the rest mass energy. If all four components of Σp_μ are conserved in any one inertial frame, they are conserved in all inertial frames: conservation of energy and momentum is Lorentz covariant. It is a matter of everyday experience that relativistic energy and momentum are conserved.

Relativistic kinematics is a standard tool of high energy physics, and this lecture is concerned with the repertoire of tricks which make most real problems readily soluble.

First, momentum and energy make up a 4-vector so it is trivial to obtain momentum and energy measureable in one frame from momentum and energy (of the same particle) measured in another:

$$\begin{aligned} p'_x &= \frac{p_x - vE/c^2}{\sqrt{1 - v^2/c^2}} \\ p'_y &= p_y \quad ; \quad p'_z = p_z \\ E' &= \frac{E - vp_x}{\sqrt{1 - v^2/c^2}} \end{aligned} \tag{6.1}$$

where the relative velocity of the two frames is v — more accurately, S' moves with velocity v along the positive x axis in S.

So one simple trick is to choose the relative motion along a common axis of two Cartesian coordinate systems - anyone can remember (6.1) or, equivalently, (3.1). Note that momentum transverse to the relative motion of the two frames is invariant, and only the longitudinal component of momentum and the energy get mixed by Lorentz transformations.

[PROBLEM: The three components of momentum specify the direction in which a particle moves and we may easily calculate the direction in one frame from the direction in another using (6.1). But momentum also has the same direction as velocity, (5.3), so compare directions using (6.1) and the more complicated velocity transformations (5.1) and show that you get the same answer.]

Secondly, the units employed in practical calculations are always those of energy rather than momentum, energy and mass separately, and velocity is measured in units of the velocity of light. Thus

$$\begin{aligned} p'_x c &= \frac{p_x c - \frac{v}{c}E}{\sqrt{1 - v^2/c^2}} = \gamma(p_x c - \beta E) \\ E' &= \frac{E - \frac{v}{c}p_x c}{\sqrt{1 - v^2/c^2}} = \gamma(E - \beta p_x c) \end{aligned} \tag{6.2}$$

where β is the dimensionless quantity v/c and γ is the dimensionless quantity $\frac{1}{\sqrt{1-v^2/c^2}}$. All quantities appearing on the far right of (6.2) have the dimensions of energy. We measure pc in units of energy, and for shorthand write p instead of pc. This is equivalent to simplifying the equations (6.2) further by defining $c = 1$ (but this way we don't lose touch with more familiar units). Then

$$\begin{aligned} p'_x &= \gamma(p_x - \beta E) \\ E' &= \gamma(E - \beta p_x) \end{aligned} \tag{6.3}$$

where if E is measured in ergs, then p is measured in ergs, which means that pc is measured in ergs and c is set equal to unity for convenience, or that p is a convenient shorthand for pc. In such units, the relation between momentum and energy takes the form

$$p^2 - E^2 = -m^2 \tag{6.4}$$

where m is shorthand for mc^2 and is again measured in terms of energy.

We usually don't use ergs (cgs) as the unit of energy, but then we don't use joules (mks) either. For largely historical reasons, the unit of energy is the MeV, or multiples thereof. One MeV is the energy acquired by a particle carrying the electron charge in falling through a potential of 10^6 volts. The electron charge is 1.6×10^{-19} coulombs (4.8×10^{-10} esu) and so

$$\begin{aligned} 1 \text{ MeV} &= 1.6 \times 10^{-13} \text{ joules}(1.6 \times 10^{-6} \text{ ergs}) \\ 1 \text{ MeV} &= 10^3 \text{ KeV} = 10^6 \text{ eV} \\ 1 \text{ GeV} &= 10^3 \text{ MeV} = 10^9 \text{ eV} \\ 1 \text{ TeV} &= 10^3 \text{ GeV} = 10^{12} \text{ eV} \end{aligned}$$

The standard unit of momentum is MeV/c (or GeV/c) and the standard unit of mass is MeV/c^2 (or GeV/c^2). With c set equal to 1, we loosely talk of the units of E, p, m all being MeV (or GeV).

EXAMPLES:

(i) The electron mass is 9.1×10^{-28}gm.
Then

$$\begin{aligned} m_e c^2 &= 9.1 \times 10^{-28} \times (3 \times 10^{10})^2 \quad \text{ergs} \\ &= \frac{9.1 \times 10^{-28} \times (3 \times 10^{10})^2}{1.6 \times 10^{-6}} \quad \text{MeV} \\ m_e &= 0.51 \quad \text{MeV}/c^2 \end{aligned}$$

(ii) An inverse calculation
The mass of the proton is 938 MeV/c^2

$$\begin{aligned} m_p c^2 &= 938 \times 1.6 \times 10^{-6} \quad \text{ergs} \\ m_p &= \frac{938 \times 1.6 \times 10^{-6}}{(3 \times 10^{10})^2} \quad \text{gm} \\ &= 1.67 \times 10^{-24} \quad \text{gm} \end{aligned}$$

(iii) A particle has momentum 17 GeV/c

Then

$$pc = 17 \times 1.6 \times 10^{-3} \quad \text{ergs}$$

$$p = \frac{17 \times 1.6 \times 10^{-3}}{3 \times 10^{10}} \quad \text{cgs units}$$

$$= 9.07 \times 10^{-13} \quad \text{cgs units} \quad (1 \text{ gm } \times 1 \text{ cm sec}^{-1})$$

(iv) A particle with unit charge (for example, an electron) is moving in a circle at right angles to a magnetic field B of 10 kG (1 tesla). The radius of the orbit is 100m. What is the momentum?

The condition for circular motion is

$$\begin{aligned} p &= Bev \text{(SI units)} \\ &= 1 \times 1.6 \times 10^{-19} \times 100 \\ &= 1.6 \times 10^{-17} \text{ mks units} \\ pc &= 1.6 \times 10^{-17} \times 3 \times 10^{8} \text{ joules} \\ &= 4.6 \times 10^{-9} \text{ joules} \\ &= 30 \text{ GeV} \end{aligned}$$

$$\begin{aligned} p &= \tfrac{1}{c}Bev \quad \text{(Gaussian cgs units)} \\ &= \frac{10^4 4.8 \times 10^{-10} \times 10^4}{3 \times 10^{10}} \\ &= 1.6 \times 10^{-12} \text{ cgs units} \\ pc &= 1.6 \times 10^{-12} \times 3 \times 10^{10} \text{ ergs} \\ &= 4.8 \times 10^{-2} \text{ ergs} \\ &= 30 \text{ GeV} \end{aligned}$$

$$p = 30 \text{ GeV}/c$$

[PROBLEM: A π° meson at rest decays into two γ's (photons, of rest mass zero) each of energy 68 MeV. What is the mass of the π° in units of (a) MeV/c^2 (b) grams.]

Thirdly, some useful quantities and relations.

In its rest frame, the 4-momentum of a particle is $(0, im_0)$: zero momentum, energy m_0. In some other frame in which the particle has velocity $\mathbf{v}$, eqs.(6.3) yield

$$\mathbf{p} = \gamma\beta m_0$$
$$E = \gamma m_0$$

Therefore in any frame

$$\frac{p}{E} = \beta \qquad \frac{E}{m_0} = \gamma \tag{6.5}$$

(and of course the relation $p^2 - E^2 = -m_0^2$ is satisfied). The measured lifetime of a particle having proper lifetime τ is thus

$$t = \gamma\tau = \frac{E}{m_0}\tau$$

and the mean distance travelled is

$$\lambda = vt = v\gamma\tau = \beta\gamma c\tau = \frac{p}{m_0}c\tau$$

The relations (6.5), derived for a single particle, have a more general applicability. If we have a group of particles with momenta and energies $\mathbf{p}_i$, E_i then

$$\beta_{\text{cm}} = \frac{|\Sigma \mathbf{p}_i|}{\Sigma E_i} \tag{6.6}$$

and

$$\gamma_{\text{cm}} = \frac{\Sigma E_i}{E_{\text{cm}}}$$

where $\beta_{\rm cm}$ is the velocity of the centre of mass of the particles in the frame in which the momenta and energies are $\mathbf{p}_i$, E_i, and $E_{\rm cm}$ is the total energy of those particles in the centre of mass frame. This is easily seen in physical terms by inventing a particle of mass $\mathcal{M}$ which decays into the observed group of particles, with conservation of energy and momentum. The total energy of the particle is ΣE_i and the total momentum $\Sigma \mathbf{p}_i$; the rest frame of the imagined particle is the centre of mass frame of the group, for in that frame the total momentum of the observed particles is zero:

$$\sum_i \mathbf{p}_i^{\rm cm} = 0$$

It would be less ambiguous to call it the centre of momentum frame—it should be obvious that this frame is the frame of the centre of inertial mass and not the centre of proper mass.

[PROBLEM: If it isn't obvious, prove it.]

The summed 4-momenta of a group of particles constitutes a 4-vector

$$\sum_i p_\mu^i$$

and so the quantity

$$s = (\sum_i E_i)^2 - (\sum_i \mathbf{p}_i)^2 \qquad (6.7)$$

is an invariant and is equal to the square of the energy (total energy—the sum of rest mass energies and kinetic energies) in the centre of mass frame. The quantity $E_{\rm cm} = \sqrt{s}$ is the mass of a group of particles—if the group is produced by the decay of a particle of mass $\mathcal{M}$ then $\mathcal{M} = \sqrt{s}$. The invariant s is enormously useful. A good rule for solving problems in relativity is to work with invariants whenever possible, and a great many problems involving particle kinematics are best tackled by starting in the centre of mass frames at least at the thinking stage.

[PROBLEM: Show that for any group of particles with arbitrary momenta and energy (subject to the condition that they all have $\beta \leq 1$) there exists a frame in which $\sum_i \mathbf{p}_i = 0$, which has velocity $\leq c$ in any other inertial frame. (Compare with timelike invariant intervals). Show also that the velocity of this frame is given by (6.6).]

Some simple EXAMPLES: (i) A particle with proper mass m_1 is incident upon a second particle with proper mass m_2, which is at rest. Calculate the centre of mass energy and the velocity of the centre of mass.

The total momentum of the system of two particles is p_1, and the total energy is $E_1 + m_2$. The velocity of the centre of mass is therefore

$$\beta_{\rm cm} = \frac{p_1}{E_1 + m_2}$$

[PROBLEM: Check this by boosting into a frame in which the two particles have equal and opposite momentum, using (6.3).]

The centre of mass energy is given by

$$\begin{aligned} E_{\rm cm}^2 &= (E_1 + m_2)^2 - p_1^2 \\ &= m_1^2 + m_2^2 + 2m_2 E_1 \\ E_{\rm cm} &= (m_1^2 + m_2^2 + 2m_2 E_1)^{\frac{1}{2}} \end{aligned}$$

Thus the available energy for making new particles increases only as the square root of the projectile energy when working with a stationary target. This is one reason why machines for studying particle physics at the highest energies are now colliding beam devices.

[PROBLEM: The particle accelerator HERA, under construction at DESY in Hamburg, will collide 30 GeV/c electrons head-on with 800 GeV/c protons. Calculate the available energy in the centre of mass, the velocity of the centre of mass in the laboratory and the momentum of either particle in the centre of mass. (Answers: 310 GeV, 0.93c, 155 GeV/c)]

[PROBLEM: At relatively low energy in the centre of mass, elastic scattering of π^+ mesons on protons is dominated by formation of the particle Δ^{++}, followed by its decay back into $\pi^+ p$. The cross section exhibits the Δ^{++} peak at a π^+ momentum of 0.3 GeV/c in the laboratory, in which the hydrogen target is at rest. Calculate the mass of Δ^{++} ($m_{\pi^+} = 139.6$ MeV/c^2, $m_p = 938$ MeV/c^2).]

(ii) A particle of (proper) mass M decays into two particles each of mass m. Find the momenta in the centre of mass frame of the decay products.

The decay products must have equal and opposite momenta in the centre of mass, and because their masses are identical, they have the same energy in the centre of mass. Then

$$M = 2\sqrt{p^2 + m^2}$$

and the magnitude of the centre of mass momentum of each is given by

$$p = \sqrt{\frac{M^2}{4} - m^2}$$

In general the two particles will have different masses. For a two particle decay the momenta must still be equal and opposite, but the energies are no longer equal. Rather,

$$M = \sqrt{p^2 + m_1^2} + \sqrt{p^2 + m_2^2}$$

whence

$$p^2 = \frac{M^4 + m_1^4 + m_2^4 - 2M^2 m_1^2 - 2M^2 m_2^2 - 2m_1^2 m_2^2}{4M^2}$$

This is another relation which is extremely useful.

[PROBLEM: Prove it.]

[PROBLEM: Find the energy of each particle in terms of the three masses M, m_1, m_2.]

(iii) Suppose two particles have back-to-back momenta $p_{\rm cm}$ and energies $E^1_{\rm cm}$, $E^2_{\rm cm}$, and their relative momentum vector makes an angle θ with the direction of motion of the centre of mass (the direction opposite to the direction of motion of the laboratory in the centre of mass frame. What are their angles, measured with respect to the direction of motion of the centre of mass system, in the laboratory?

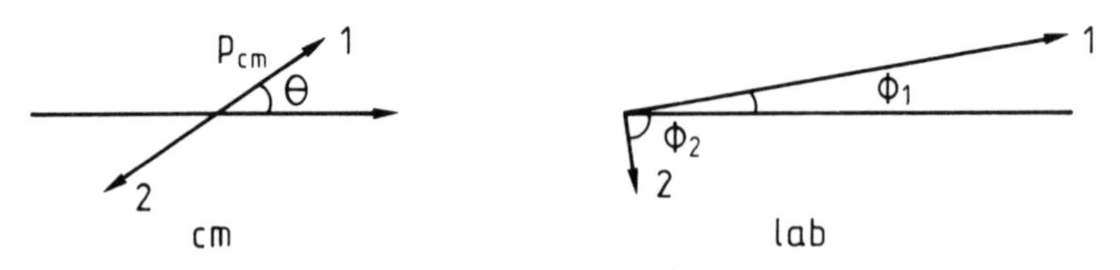

We don't use the velocities—such an approach will work but it is slow and there are more opportunities for mistakes. Calculate the momentum vectors of the two particles in the laboratory as a function of the parameter θ.

$$p^1_{x\text{lab}} = (p^1_{x\text{cm}} + \beta E^1_{\text{cm}})\gamma = (p_{\text{cm}}\cos\theta + \beta E^1_{\text{cm}})\gamma$$
$$p^2_{x\text{lab}} = (p^2_{x\text{cm}} + \beta E^2_{\text{cm}})\gamma = (-p_{\text{cm}}\cos\theta + \beta E^2_{\text{cm}})\gamma$$

The transverse momentum of each particle is invariant and is equal to $p_{\text{cm}}\sin\theta$. So

$$\tan\phi_1 = \frac{p_{\text{cm}}\sin\theta}{(p_{\text{cm}}\cos\theta + \beta E^1_{\text{cm}})\gamma} = \frac{\sin\theta}{(\cos\theta + \beta/\beta^1_{\text{cm}})\gamma}$$
$$\tan\phi_2 = \frac{p_{\text{cm}}\sin\theta}{(-p_{\text{cm}}\cos\theta + \beta E^2_{\text{cm}})\gamma} = \frac{\sin\theta}{(-\cos\theta + \beta/\beta^2_{\text{cm}})\gamma}$$

where β and γ characterise the centre of mass motion relative to the laboratory.

The utility of such an approach in particle decay processes is obvious. It can often be useful to set up other problems in the centre of mass system and then transform to the laboratory rather than to solve the equations of conservation of momentum and energy directly in the laboratory.

The utility of the tricks of the trade can only be appreciated by working out problems ....

[PROBLEM: Consider $\pi^+ p$ elastic scattering. Show that if the proton, initially at rest in the laboratory, is scattered forwards, at 0° to the incident beam, then the π^+ emerges backwards in the laboratory regardless of the incident pion energy. ($m^+_\pi = 139.6$ MeV/c^2; $m_p = 938$ MeV/c^2).]

[PROBLEM: The vector meson ω° decays into $\pi^+\pi^-\pi^\circ$. Calculate the maximum and minimum momentum of any pion
(a) In the centre of mass frame of the ω°
(b) In the laboratory, if the ω° momentum is 3 GeV/c ($m_\omega = 0.783$ GeV/c^2; take all three pions to be of equal mass).]

[PROBLEM: A 50 GeV neutrino is scattered by an electron at rest in the laboratory. Calculate the angle to the neutrino beam at which the electron emerges as a function of the scattering angle in the centre of mass. Evaluate the angle for $\theta_{\text{cm}} = 90°$ (Answer: 0.26°).]

[PROBLEM: The Υ (a bound state of $b - \bar{b}$ quarks) has a mass of 9.46 GeV/c^2. It is produced at $e^+ - e^-$ colliding beam machines through $e^+e^- \to \Upsilon$ at an energy of 4.73 GeV in each beam. If electrons at rest were used as a target, what energy positrons would be needed?]

[PROBLEM: Stationary protons are bombarded with a beam of protons which have been accelerated through a potential difference V volts in the laboratory. What is the minimum value of V for the production of antiprotons through the reaction

$$p + p \to p + p + p + \bar{p}$$

to be energetically possible?

A heavy nucleus in which the nucleons have Fermi momentum ~ 150 MeV/c is used as a target instead. Estimate the minimum value of V now required.]

[PROBLEM: Calculate the minimum pion laboratory momentum necessary for the reaction

$$\pi^- + p \to \Lambda^\circ + K^\circ$$

($m_\Lambda = 1.115$ GeV/c^2; $m_{K^\circ} = 0.498$ GeV/c^2).]

[PROBLEM: An accelerator produces a burst of K^+ mesons which (after momentum selection) have a momentum of 10 GeV/c. If they must travel a distance of 100m from the production point to a detector, how many decay before getting there? ($m_{K^+} = 496$ MeV/c^2, K^+ proper mean lifetime 1.2×10^{-8} s).]

[PROBLEM: Unless they have been filtered out, π^+ and protons will predominate over the K^+ in a burst of secondary particles travelling down a momentum-selected beam line as in the previous question. Find the time intervals (in the laboratory) separating the arrival at the detector of the three types of particle.]

[PROBLEM: Classically, when a proton scatters from another at rest in the laboratory, the opening angle between the two scattered protons is always 90°. Show that this angle is $\leq 90°$ when relativistic kinematics is used. For what configuration is the angle 90°?]

[PROBLEM: A K° of momentum 3 GeV/c decays into $\pi^+\pi^-$. Find the maximum opening angle between the two pions in the laboratory. (The answer is close to the result obtained for decay at 90° to the line of flight in the centre of mass, but this is not the configuration. You should find the decay angle is 41° in the centre of mass for the maximum opening angle in the laboratory.)]

[PROBLEM: The neutral pion, π°, decays into two photons. In the π° rest frame the decay is isotropic. If the π° has energy E in the laboratory, find the energy spectrum of the photons in the laboratory (the number of photons in unit energy interval, regardless of the angle made with the line of flight). The π° was discovered through the observation of just such a spectrum. (The calculation is easy but it takes a lot of thought to see how to do it. Start in the π° rest frame and calculate the photon energies in the laboratory.)]

[PROBLEM: $\pi^- p$ elastic scattering may be identified by demanding signals in coincidence from two small counters, one on each side of the incident π^- beam. Find the relation betweeen the angles at which the two counters must be placed for incident π^- momenta of a) 1 GeV/c b) 3 GeV/c.]

[PROBLEM: Among the debris from a high energy interaction is a neutral particle which decays into $K^+\pi^-$ a distance of 0.9mm from the production vertex. The K^+ has momentum 10.122 GeV/c, the π^- 1.047 GeV/c. These two particles are coplanar with the line of flight and at angles of 2.79° and 28.13° to the line of flight, on opposite sides. Calculate the mass of the neutral particle, the angle between the K^+ and the line of flight in the centre of mass of the neutral particle, and the proper lifetime of this particular specimen. (Answers: 1.868 GeV/c^2, 35°, 5×10^{-13}s)]

[PROBLEM: Here are data from 5 events consistent with the process $\Omega^- \to \Lambda K^-$, where the Λ is clearly identified and the negative particle is presumed to be a kaon:

$p(\Lambda)$	$p(K)$	$p(\Omega^-)$	$M(\Lambda K^-)$
2.0	0.82	2.8	1.670
1.1	0.45	1.5	1.665
2.1	0.85	2.9	1.670
1.7	0.79	2.5	1.669
1.6	0.77	2.3	1.667

Could these events plausibly be examples of $\Xi^- \to \Lambda\pi^-$? Start in the centre of mass of the decaying object and calculate for each event $p(\Lambda)$, $p(K)$ as a function of the centre of mass angle between the K^- and the Ω^- line of flight and hence find the value of this angle for each event. Then repeat the exercise assuming the events are in reality Ξ^-

decay and finally calculate for each putative Ξ^- event the mass of the Λ and the charged particle when the latter is assumed to be a kaon. Take $M_{\Omega^-} = 1.672$, $M_{\Xi^-} = 1.321$, $M_{\Lambda^\circ} = 1.116$, $M_{K^-} = 0.493$, $M_{\pi^-} = 0.140$. All units are GeV. (The Ξ^- will decay isotropically in its own rest frame. This problem is best tackled with a computer or a programmable calculator).]

[PROBLEM: A particle travels a distance ℓ in the laboratory before decaying. Visible tracks of the decay products can be extrapolated back and the perpendicular distance d between the line of flight of a decay product and the point at which the primary particle was produced can be constructed. Show that to a good approximation the distance d depends only on the proper lifetime of the primary particle and the angle in the centre of mass of the primary at which the secondary was produced, and in particular is independent of the value of γ for the primary. (Assume that $\beta \sim 1$ for both the primary in the laboratory frame and for the secondary in the primary rest frame. Start in the primary rest frame.) If secondaries are produced isotropically in the primary rest frame, find the relation between the proper mean life τ of the primary and the value of d averaged over many different secondaries from many different primaries (Answer: $\tau \simeq \frac{2}{\pi}\frac{<d>}{c}$). Is d exponentially distributed? (No.)]

Lecture 7 SOME ASPECTS OF ACCELERATION

Even in 1984 the rumour persists that special relativity is incapable of dealing with accelerations. It is a popular belief that in order to discuss accelerations general relativity is required, and an even more popular belief that general relativity is too difficult to be understood by more than a handful of specialists. The rumour and the popular beliefs are utterly false. Most of physics is concerned with objects changing their states of motion, undergoing accelerations, and special relativity would be useless were it incapable of dealing with these phenomena. Special relativity deals with the phenomena of acceleration perfectly.

We may distinguish several aspects:

(1) The physics of an idealised point particle undergoing acceleration
(2) The physics of an extended and perhaps complicated system undergoing acceleration
(3) The physics obtaining in an accelerated laboratory.

In the first two cases we already understand what is going on. Under the assumptions of special relativity, we describe the physics by equations which are covariant under the Lorentz transformations. These equations contain the behaviour of a system at rest in the laboratory, which in the second case involves the acceleration of at least parts of the system, and automatically include the behaviour of the system when moving with some arbitrary velocity with respect to the laboratory system. If we have all the necessary equations and they are Lorentz covariant, then we can solve the latter problem either directly from the equations, given the acceleration and velocity, or by considering a system at rest with the appropriate acceleration and then boosting to a frame in which it has the desired velocity. Conversely, many aspects of the physics obtaining in an accelerated laboratory can be handled by working out what happens in a single inertial frame, and then transforming to successive instantaneous rest frames of the accelerated laboratory. The point is that the physics is labelled by a sequence of events at space-time points and the relation between the coordinates in two different frames is given by the Lorentz transformations, which are only velocity dependent. The same is true of Newtonian physics: the Galilean transformations are velocity dependent only. The physical assumptions are merely that covariant equations embodying the effects of acceleration plus the Lorentz transformations tell the whole story. True or false? If we merely knew that the Lorentz transformations ensured that the speed of light in any inertial frame is a universal constant, we might wonder. But we know that the Lorentz transformations leave Maxwell's equations and the Lorentz force law in the same form and with the same numerical content, so the recipe should (and does) work at least for electromagnetic problems.

For example, the covariant equation of motion for an electron in a general electromagnetic field can be written

$$\frac{dp_\mu}{d\tau} = F_\mu \tag{7.1}$$

where F_μ is the covariant Lorentz force, and in an inertial frame in which the electromagnetic field consists of a constant electric field $\mathbf{E}$ this becomes

$$\frac{d\mathbf{p}}{dt} = e\mathbf{E} \quad ; \quad \frac{dE}{dt} = e\mathbf{v} \cdot \mathbf{E} \tag{7.2}$$

where $\mathbf{p}$ is the momentum of the electron ($\gamma m_0 \mathbf{v}$) and E the energy ($\gamma m_0 c^2$). Maxwell's equations tell us all about the physics of an arbitrarily accelerated electron. If they didn't, our particle accelerators would not work.

[PROBLEM: Electrons at SLAC gain 20 GeV over 3 km. Assuming they are accelerated in a constant electric field, calculate the laboratory acceleration as a function of time.]

[PROBLEM: Protons accelerated in the synchrotrons at CERN and Brookhaven reach an energy $\sim$ 30 GeV in $\sim$ 1s, moving in circular orbits determined by iron-cored magnets. They experience the rf accelerating fields for $\sim 10^{-2}$ of each orbit. Estimate the peak tangential accelerations and calculate the maximum radial acceleration (in the laboratory frame).]

An example of an explicitly calculable system in the second category is provided by a classical atom, perhaps singly ionised helium. Apply to it an electric field sufficiently weak that the properties (calculated from electrostatics and mechanics) are not significantly changed when it is at rest, and follow the behaviour of the electron as the atom gently accelerates to very high speeds, using the full set of Maxwell's equations and relativistic mechanics. Explicit calculation will show that the period is increased by a factor γ and the longitudinal extent of the orbit is shrunk by the same factor, where $\gamma = (1 - v^2/c^2)^{-1}$ and v is the instantaneous velocity relative to an observer in some specified inertial frame. (This kind of system was considered by Larmor and more recently by Bell; see references at the end of Lecture 1. A realistic example is discussed by J.D. Lawson, *Nature* **218** 430 (1968)). However, knowing the accelerations are sufficiently gentle not to distort (or break up) the atom, and that the equations of motion are covariant, we could reach these conclusions at once from the Lorentz transformations, without even making an explicit calculation of the motion from the dynamical equations. A classical treatment of an atom is only valid for orbitals of very high principal quantum number, and they are only too easy to break up, but atomic structure is described with enormous precision by the (relativistic) theory of quantum electrodynamics.

[PROBLEM: Estimate the instantaneous velocity and radial acceleration of an electron a) bound in a hydrogen atom b) in the lead K-shell.]

It is worth looking again at the machinery for discussing accelerations in terms of covariant equations of motion.

Consider the motion of a particle in some specified inertial frame S in which the coordinates of the particle are x_μ. We define the 4-velocity of the particle to be

$$\lim_{\Delta t \to 0} \frac{\Delta x_\mu}{\sqrt{\Delta t^2 - \Delta \mathbf{x}^2}} = \frac{dx_\mu}{d\tau} \quad \text{where } \Delta \mathbf{x} = \mathbf{v}\Delta t (\text{and } c = 1)$$

If the particle is not accelerated, a single Lorentz transformation determined by $\mathbf{v}$ always takes us from S to the particle frame. The interval $\Delta\tau = \sqrt{\Delta t^2 - \Delta \mathbf{x}^2}$, constructed from measurements made in the frame S, is an invariant and is an interval of proper time in an inertial frame momentarily moving with the particle — if the equations governing the ticks of clocks in the particle rest frame are covariant, then those clocks keep the time τ.

We may in principle measure

$$\lim_{\Delta t \to 0} \Delta \left\{ \frac{\Delta x_\mu}{\sqrt{\Delta t^2 - \Delta \mathbf{x}^2}} \right\} \Big/ \sqrt{\Delta t^2 - \Delta \mathbf{x}^2} = \frac{d^2 x_\mu}{d\tau^2} \tag{7.3}$$

which is the 4-acceleration. Again, all measurements may be made in the inertial frame S and in this frame the instantaneous velocity is given by

$$\lim_{\Delta t \to 0} \frac{\Delta \mathbf{x}}{\Delta t}.$$

The quantity $\sqrt{\Delta t^2 - \Delta \mathbf{x}^2}$ is still Lorentz invariant and is the proper time kept by real clocks in an inertial frame moving with the instantaneous velocity of the particle. In such a frame, which has constant velocity, the particle is still accelerated, for the 4-acceleration

$$a_\mu = \frac{d^2 x_\mu}{d\tau^2} \tag{7.4}$$

is by construction a 4-vector and may be transformed into that frame which has instantaneously the same velocity as the test particle. We may thus track the particle by successive Lorentz transformations to the instantaneous rest frame.

Now we work backwards. A particle (or a more complicated system such as a space vehicle containing a mass on a spring) is at rest momentarily in a given inertial frame S', but subject to an acceleration. Then $\frac{d\mathbf{x}'}{d\tau} = 0$ but $\mathbf{a} = \frac{d^2\mathbf{x}'}{d\tau^2} \neq 0$ — it is the acceleration in the frame where the particle is at rest, the proper acceleration. The quantity

$$\frac{dt'}{d\tau} = 1 \quad \text{and} \quad \frac{d^2 t'}{d\tau^2} = 0 \qquad \text{[PROBLEM : Prove it.]}$$

Thus

$$a'_\mu = (\mathbf{a}; 0) \quad \text{while} \quad V'_\mu = (0; c) \tag{7.5}$$

where again the construction of a'_μ makes it a 4-vector — which can now be transformed from the inertial frame S' moving with the particle to any other inertial frame. Thus if we understand the physics of an accelerated system in a frame in which its velocity is instantaneously zero, we can calculate the behaviour as measured as a function of time in any specified inertial frame.

Let us transform a_μ from the comoving inertial frame (not a single inertial frame but the family of inertial frames in which the accelerated system is at rest) to a specified inertial frame S in which the comoving inertial frame has velocity $\mathbf{v}$, in the same direction as the acceleration.

$$a_x = \frac{a}{\sqrt{1 - v^2/c^2}} = \frac{d^2 x}{d\tau^2} \qquad a_4 = \frac{iav/c}{\sqrt{1 - v^2/c^2}}; \quad \frac{d^2 t}{d\tau^2} = \frac{av/c^2}{\sqrt{1 - v^2/c^2}} \tag{7.6}$$

(Note that $V_x = \frac{v}{\sqrt{1-v^2/c^2}}$, $V_4 = \frac{ic}{\sqrt{1-v^2/c^2}}$ so $a_\mu V_\mu = 0$)

Remember that a_x is the x component of the 4-acceleration, $\frac{d^2 x}{d\tau^2}$ whereas the acceleration is more familiarly defined to be $\frac{d^2 x}{dt^2}$. We may find the familiar acceleration in the following way:

$$\begin{aligned}\frac{d^2 x}{d\tau^2} &= \frac{d}{d\tau}\left[\frac{dx}{d\tau}\right] = \frac{d}{d\tau}\left[\frac{dx}{dt}\frac{dt}{d\tau}\right] = \frac{d^2 x}{dt^2}\left(\frac{dt}{d\tau}\right)^2 + \frac{dx}{dt}\frac{d^2 t}{d\tau^2} \\ &= \frac{d^2 x}{dt^2}\frac{1}{1 - v^2/c^2} + v\frac{d^2 t}{d\tau^2}\end{aligned}$$

$$\frac{1}{1 - v^2/c^2}\frac{d^2 x}{dt^2} = \frac{a}{\sqrt{1 - v^2/c^2}} - \frac{av^2/c^2}{\sqrt{1 - v^2/c^2}}$$

$$\frac{d^2 x}{dt^2} = a(1 - v^2/c^2)^{3/2} \tag{7.7}$$

[PROBLEM: What is the analogue of (7.7) for acceleration transverse to $\mathbf{v}$?]

For a constant proper acceleration, the laboratory acceleration tails off as $v \to c$: compare eqn. (5.8).

A covariant equation of motion takes the form

$$m_0 \frac{d^2 x_\mu}{d\tau^2} = F_\mu$$

and in the rest frame F_μ must take the form $(\mathbf{F},0)$. This could also be written as

$$\frac{dp_\mu}{d\tau} = F_\mu$$

whence

$$\frac{d\mathbf{p}}{d\tau} = \frac{\mathbf{F}}{\sqrt{1 - v^2/c^2}} \quad \text{or} \quad \frac{d\mathbf{p}}{dt} = \mathbf{F} \quad ; \quad \frac{dE}{dt} = \mathbf{v} \,.\, \mathbf{F} \tag{7.8}$$

from which (7.7) is rapidly recovered.

The whole description is internally consistent and its applicability is validated by success of covariant equations of motion in describing systems in which particles have very high acceleration. In addition to those already considered in preceding problems, particle interactions at high energy involve even higher accelerations, and relativistic momentum and energy are still conserved.

[PROBLEM: Estimate the acceleration experienced in the laboratory by a proton elastically scattered by another through $\sim 90°$ in the centre of mass. Take the proton energy to be 200 GeV in the laboratory.]

[PROBLEM: Estimate the acceleration experienced by an electron or positron in elastic backscattering, the experiment being performed with colliding beams of e^+e^- having energies of 20 GeV each, as in the storage ring PETRA at DESY in Hamburg.]

We may finally note that the proper acceleration which we have defined coincides in all respects with the ordinary non-relativistic definition of acceleration as measured, for example by the extension of a spring.

By now you should be (almost) convinced that special relativity is entirely capable of application to situations involving acceleration. The formalism of general relativity provides an elegant way of handling the problem of the physics obtaining in an accelerated frame, but is quite unnecessary for that purpose. General relativity is a theory of gravitation, and is certainly the most convenient and concise way of handling that theory. Even in this application, special relativity is not defficient: provided that Lorentz covariant equations can be written for gravitational problems (they can, but have some curious and unexpected features) then special relativity may be used, although in most cases it is doing things the hard way. [See M.G. Bowler, Gravitation and Relativity, Pergamon 1976].

[PROBLEM: In any inertial frame, light travels in a straight line with constant velocity. Consider the trajectory followed by light in an accelerated laboratory.]

A particularly simple example of accelerated motion obtains for constant proper acceleration, for example an electron in a constant uniform electric field, or a space vehicle with its engines servoed to an onboard accelerometer such that the accelerometer always reads 1 g for comfort. The equation of motion, in a specified inertial frame, is

$$\frac{d^2 x}{dt^2} = a(1 - v^2/c^2)^{3/2} \tag{7.9}$$

and it is trivial to show that the velocity is asymptotic to c. This example is known as hyperbolic motion, because the trajectory in the (x,t) plane is an hyperbola.

[PROBLEM: Prove it. Then consider the following problem, a relativistic Achilles and the Tortoise. A starship leaves earth at a constant proper acceleration of 1 g. Houston communicates with the starship by radio—show that after a certain time no messages sent from Houston reach the ship, although messages from the ship continue to reach Houston. Find the critical time. (The answer is $\sim$ 1 year. This is an example of a one-way event horizon in special relativity.) Hint: You can calculate explicitly, but drawing the trajectory of the starship on a space-time diagram is a great help.]

[PROBLEM: A toy version of quantum chromodynamics, which binds quarks into hadrons, represents mesons as quark-antiquark pairs, each pair joined by a (colour) string of constant tension; that is, the force between the quarks is independent of their separation. (Compare with the force between the plates of a parallel plate capacitor, one dimensional electrostatics.) The model takes its simplest form when all motions are confined to one space dimension. The equation of motion of the quarks in this single dimension is

$$\frac{dp}{dt} = \pm\alpha$$

where α is an invariant.

(i) Show that this equation of motion is covariant with respect to Lorentz transformations in the single space dimension a) for massless quarks b) for quarks with rest mass μ. (Abstract mathematics will not be much help)

(ii) A meson, consisting of two quarks of equal mass joined by a string, has mass m. Calculate the momenta of the quarks, in the rest frame of the meson, at the moment when the length of the string is zero, and find the maximum excursion of the quarks. Draw the quark world lines in x, t first in the meson rest frame and secondly in a frame where the meson is moving with velocity v. (You can see you could get this result two ways). Carry out this program first for massless quarks, and then for quarks of mass μ.

(iii) In e^+e^- annihilation a $q\bar{q}$ pair may be produced, flying apart back to back in the centre of mass system. A string stretches between them. This string is cut in many places by creation of further $q\bar{q}$ pairs. Consider two such pairs, produced at (x_1, t_1) and (x_2, t_2). The quark from one pair is linked to the antiquark from the other by the length of string snipped off, forming a meson. Calculate the mass squared of the meson and its rapidity y, defined by $y = \frac{1}{2}\ln\left\{\frac{E+p}{E-p}\right\}$, in terms of the pair creation coordinates. Assume massless quarks. (Note: This model is relativistic, but is not quantised—all meson masses are possible).]

There is an ancient problem of accelerated motion in special relativity which is usually known as the clock paradox or the twin paradox. These are misnomers, for there is nothing paradoxical in the problem at all, rather the results seem to many counter-intuitive, and the problem can be posed in homely terms. Consequently it has attracted a disproportionate amount of attention, and here we go again. A clock is at rest in a given inertial frame. An identical clock is accelerated away (sufficiently gently that it doesn't break under its own weight), achieves very high speeds and is eventually brought back to rest beside the original clock. The problem is to find the final readings of the two clocks and the reputed paradox is that they differ, the travelling clock recording a lesser elapsed time.

The analysis is essentially that employed between eqs. (7.3) and (7.4). We first remove inessential complications by requiring that the clocks are sufficiently robust that acceleration will not break them, and further that their rate is not affected by acceleration. We may check this for any particular clock by subjecting it to acceleration in the rest frame of an identical clock and comparing the rates (during periods of acceleration sufficiently short that the relative velocity remains negligible.) Acceptable clocks, when accelerated, then keep the same time as identical clocks in the instantaneous rest frame.

We now track the accelerated clock from just one inertial frame, that in which it was originally at rest and in which a comparison clock remains at rest. In this frame we measure an incremental distance Δx which occurs in a time interval Δt. The quantity

$$\Delta t^2 - \Delta x^2 \qquad \text{(units } c = 1\text{)}$$

is an invariant and is the interval of proper time which would be measured by a real clock at rest in a frame instantaneously moving at the same speed as the accelerated clock. But the accelerated clock rate matches the rate of a clock in that instantaneously comoving frame, and therefore

$$\Delta\tau = \sqrt{\Delta t^2 - \Delta x^2}$$

where Δt, Δx are measured in an inertial frame and $\Delta\tau$ is the corresponding (proper) time interval measured on the accelerated clock. We can therefore write

$$d\tau = \sqrt{1 - v^2/c^2}\,dt \tag{7.10}$$

where dt is measured in an inertial frame in which the instantaneous velocity of the accelerated clock is v. Now provided we track the accelerated clock from just one inertial frame, we may integrate (7.10) to find

$$\tau = \int \sqrt{1 - v^2/c^2}\,dt \le t \tag{7.11}$$

The supposedly paradoxical element enters as follows. It is argued that whereas the accelerated clock A recedes with increasing velocity from the inertial frame clock B, and eventually approaches it again, slowing down and coming to rest, from the point of view of B clock A recedes with increasing velocity .... Therefore use (7.11) to relate A to B and B to A, yielding contradictory results. The resolution of this apparent paradox is obvious. Equation (7.11) relates time recorded in a given inertial frame to the time recorded on an arbitrarily moving clock which changes its inertial frame. Equation (7.10) is generally true, but can only be integrated to yield (7.11) if t is always measured in the same inertial frame. The fact that the accelerated clock changes its inertial frame while the other does not removes the spurious symmetry between the two. The symmetry is manifestly absent, because accelerometers work.

The world lines of the two clocks, in the rest frame of the inertial frame clock B, would be

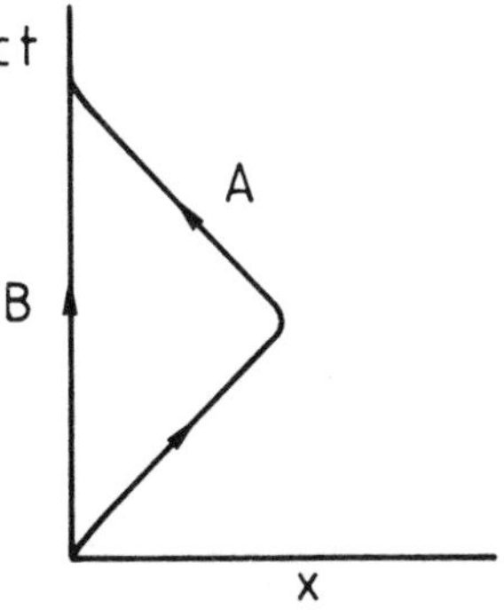

With the recipe that the length of A's world line is formed from segments $\sqrt{\Delta t^2 - \Delta x^2}$, it is obvious that $\tau_A \le \tau_B$ and it is also obvious that this is no more paradoxical than

saying that a man travelling from London to Paris via Vienna covers more ground than a man travelling direct.

[PROBLEM: (The twin problem plus hyperbolic motion) Castor and Pollux are the popular identical twins. Castor is an astronaut who makes a trip to Arcturus in a comfortable starship whose engines always give him a proper acceleration of 1 g. He accelerates to the halfway point, turns his ship and decelerates so that he reaches Arcturus at essentially zero velocity. The return trip is accomplished in like manner. Arcturus is 40 light years away: how much elapsed time is measured by (i) clocks at starbase where Pollux remains (ii) identical clocks aboard Castor's ship? (Answer: 84, 14.6 years)]

It should be clear that relative times in the clock problem depend only on the velocity v in eq. (7.11). The role of acceleration is to introduce change of inertial frame and to determine v as a function of t, but does not enter explicitly. (Acceleration is of course necessary to bring the two clocks physically together again). An analogue of the clock problem, containing the same supposedly paradoxical elements, can be constructed without any acceleration at all. It involves three clocks and the transfer of information rather than the return of a travelling clock. The world lines of three clocks are

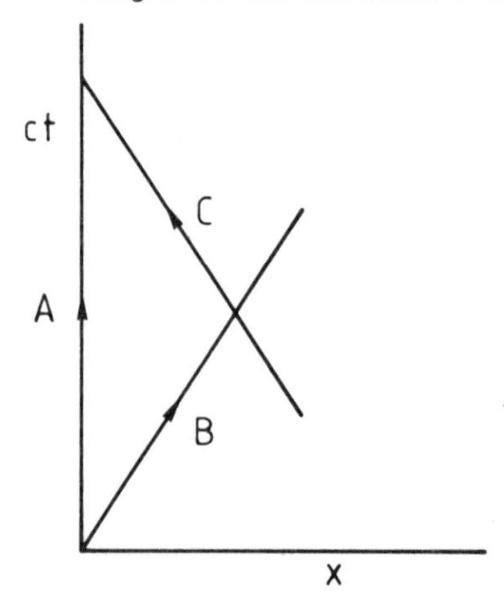

drawn in A's rest frame. In A's rest frame, B passes A at velocity v and they synchronise clocks at $t_A = t_B = 0$. At time $t_A = t_{A1}$, $t_B = t_{B1}$, C passes B heading back towards A at velocity v relative to A. C sets his clock to B's clock reading, t_{B1}, as they pass. Finally A and C compare clock readings t_{A2} and t_{C2} as they pass. Applying eq. (7.11) we have at once

$$t_{C2} = \sqrt{1 - v^2/c^2}\, t_{A2}$$

[PROBLEM: Both t_{C2} and t_{A2} are proper times, invariants. Therefore calculation of the relation between them working in either B's rest frame or C's rest frame must yield the same result. Demonstrate that this is true by explicit calculation. (Remember, the relative velocity of B and C, as measured in B's frame or C's frame, is NOT $2v$).]

We may now confront the reputedly paradoxical element for the last time. A measures a smaller time interval on B's clocks than on his own; B measures a smaller time interval on A's clocks than on his (B's) own, and C measures a smaller time interval on A's clocks than on his (C's) own. Do we not then conclude from these relative rates (which we certainly understand in terms of just two inertial frames) that $t_{C2} = \sqrt{1 - v^2/c^2}\, t_{A2}$ AND $t_{A2} = \sqrt{1 - v^2/c^2}\, t_{C2}$ — paradox? We should not. Write the

Lorentz transformations for time in the form

$$\begin{aligned} t_B &= \left(t_A - \frac{vx_A}{c^2}\right) \Big/ \sqrt{1 - v^2/c^2} \\ t_C &= \left(t_A + \frac{vx_A}{c^2}\right) \Big/ \sqrt{1 - v^2/c^2} \end{aligned} \tag{7.12}$$

or

$$\begin{aligned} t_A &= \left(t_B + \frac{vx_B}{c^2}\right) \Big/ \sqrt{1 - v^2/c^2} \\ t_A &= \left(t_C - \frac{vx_C}{c^2}\right) \Big/ \sqrt{1 - v^2/c^2} \end{aligned} \tag{7.13}$$

Thus

$$\left.\frac{\partial t_A}{\partial t_{B,C}}\right|_{x_A} = \frac{1}{\sqrt{1 - v^2/c^2}} \quad ; \quad \left.\frac{\partial t_{B,C}}{\partial t_A}\right|_{x_{B,C}} = \frac{1}{\sqrt{1 - v^2/c^2}} \tag{7.14}$$

which give us the familiar relative rates.

Remember that (7.12) and (7.13) have origins chosen such that at t_A, $x_A = 0$, t_B, x_B or t_C, x_C are zero. Thus it is most convenient to consider clocks in the B and C frame which are at their respective origins and t_A, $x_A = 0$, t_B, $x_B = 0$, t_C, $x_C = 0$ at the space time point where B and C meet. The clock A is then located in the A frame at $x_A = -X_A$, B passes the A clock at $t_A = T_{A0} = -\frac{X_A}{v}$, $t_B = T_{B0}$ and C passes the A clock at $t_A = T_{A2} = +\frac{X_A}{v}$, $t_C = T_{C2}$. The elapsed time in the A frame is thus $\frac{2X_A}{v}$. If we use (7.13) with $x_B = x_C = 0$ we at once find that

$$T_{C2} - T_{B0} = (T_{A2} - T_{A0})\sqrt{1 - v^2/c^2} \tag{7.15}$$

This is just taking the proper time intervals in B, C and relating them to A, as in equation (7.11). If instead we take eqn. (7.12) then we note that clock A is not at the origin of the coordinates of frame A, and it matters. While the rates of A frame clocks at fixed x_A relative to B or C are correctly given by (7.14), the absolute readings of corresponding B or C clocks are offset in opposite directions by an amount depending on x_A. Clocks which are synchronised in one frame are not synchronised in another. Applying (7.12)

$$\begin{aligned} T_{B0} &= \left(T_{A0} - \frac{v}{c^2}[-X_A]\right) \Big/ \sqrt{1 - v^2/c^2} \\ T_{C2} &= \left(T_{A2} + \frac{v}{c^2}[-X_A]\right) \Big/ \sqrt{1 - v^2/c^2} \end{aligned} \tag{7.16}$$

The change of inertial frame comes in through the changed sign of the term in vX_A/c^2, and

$$T_{C2} - T_{B0} = \frac{T_{A2} - T_{A0}}{\sqrt{1 - v^2/c^2}} - \frac{2vX_A}{\sqrt{1 - v^2/c^2}} \tag{7.17}$$

The first term is time dilation; contrast it with (7.15). The second term is due to the change of inertial frame from B to C. Since $T_{A2} - T_{A0} = \frac{2X_A}{v}$

$$T_{C2} - T_{B0} = (T_{A2} - T_{A0})\sqrt{1 - v^2/c^2}$$

once more.

Thus we can use time dilation in the frames of B and C, provided we work out carefully the additional effects of a change of inertial frame, but we may only use (7.11) one way round.

Referring back to eqs. (7.12), (7.14) an A clock at any fixed value of x_A runs slow relative to B, C time, by an amount independent of x_A. But the absolute reading relative to B or C depends on x_A (remember all clocks in any one frame are synchronised in that frame). If $t_A = t_B = t_C = 0$ at x_A, x_B, $x_C = 0$, then for $t_B = 0$, a clock at $-X_A$ reads $-vX_A/c^2$, while a clock at $-X_A$ reads $+vX_A/c^2$ for $t_C = 0$. We only have $t_B = t_C = 0$ simultaneously at one point in spacetime—when B and C meet. Thus in frame B an A clock at $-X_A$ lags behind an A clock at $x_A = 0$, while in frame C it leads it. This difference must be included when working from frames B and C and using time dilation for the relative clock rates. If you are travelling with B and accelerate (your velocity changing continuously in A's frame) so as to join C this difference is of course achieved continuously.

Physics is an experimental subject, and it would be nice to send Castor to Arcturus and back, but it is not practical. An equivalent experiment has been performed and the results are in excellent agreement with eq. (7.11), read the right way round.

The muon is a spin $\frac{1}{2}$ particle of mass 106 MeV/c^2. It behaves in all respects like a distinct species of heavy electron, and being heavy decays through the weak interactions

$$\mu^+ \to e^+ \nu_e \bar{\nu}_\mu \quad ; \quad \mu^- \to e^- \bar{\nu}_e \nu_\mu$$

with a proper lifetime of 2.2 μs. The leptons (and quarks) are known to behave (within quantum electrodynamics) as pointlike particles down to at least 10^{-16} cm, so muon decay should provide an exceedingly robust clock. The lifetime has been measured for muons in flight (unaccelerated): normal time dilation is exhibited. When brought to rest in a light element such as hydrogen or carbon, both μ^+ and μ^- have the same proper lifetime against decay, although the μ^- is rapidly captured into the lowest muon orbital about the carbon nucleus. The velocity in this orbital in carbon is $\sim 10^9\,\mathrm{cm\,s^{-1}}$ so γ is negligibly different from unity, but $v^2/r \sim 3 \times 10^{30}\,\mathrm{cm\,s^{-2}}$. We may be confident that under acceleration the muon behaves as an ideal clock.

In experiments designed to measure with great accuracy the gyromagnetic ratio of the muon, bursts of muons derived from the CERN PS were trapped in a storage ring of radius 7m in a magnetic field of $\sim$ 14.7 kG. The momentum of the stored muons was 3.10 GeV/c and $\gamma = 29.3$. Using eq. (7.11), the prediction of special relativity is that the bursts of muons should decay away with a mean laboratory lifetime of $29.3 \times 2.2\,\mu\mathrm{s} = 64.4\,\mu\mathrm{s}$. They did: within the accuracy of the experiment, eq. (7.11) is verified for orbiting muons with $\gamma \simeq 30$ at the 0.1% level. The laboratory acceleration of these muons was $\sim 10^{18}\,\mathrm{cm\,s^{-2}}$. [J. Bailey et al., *Nature* **268** 301 (1977)].

We may end this lecture with a more philosophical remark. The content of eq. (7.11) is that if a clock (or an astronaut) accelerates away from another which remains at rest in an inertial frame, and then returns, less time elapses for the traveller (or travelling clock). Whether or not a traveller has sufficient information to properly predict this result is irrelevant. In general he will have—his rockets are fired, his accelerometers register acceleration, his telescopes measure changing Doppler shifts from the star field and so on. However, suppose the traveller initially passes the stay-at-home at constant velocity and is later turned in a tight hyperbolic orbit about a massive star. The accelerometer won't register (although a sensitive instrument could detect the presence of a tidal force) but presumably the star field information is available. If it is demanded that the traveller is in a sealed box, of course he will not even know that he is going back. Nonetheless, eq. (7.11) still predicts the result of a comparison of the clocks when the traveller passes the stay-at-home. (There is one proviso. A gravitational potential slows down all clocks and unless the period spent in a region of strong gravitational potential is negligible in comparison with the duration of the trip, this effect would have to be included explicitly).

[PROBLEM: The final problem (of this lecture).

This is a highly artificial but interesting and informative conundrum. I first came across it when it was going the rounds of the high energy physics community after over half of a sample of the CERN Theory Division had (reputedly) got it wrong.

Two space vehicles are at rest, in line ahead, in some inertial frame. They are joined by a light string which is neither infinitely extensible nor infinitely strong.

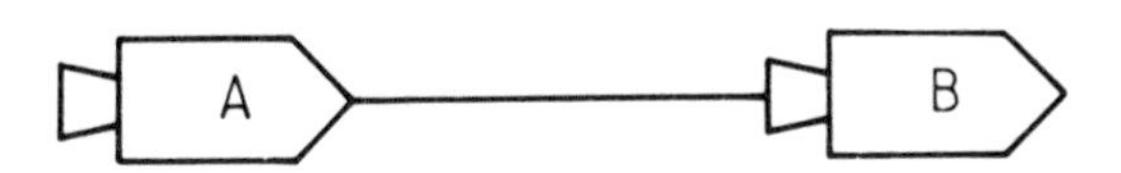

The vehicles are identical and equipped with identical accelerometers and computers programmed to fire the engines simultaneously (in the original rest frame) and impart to each vehicle a constant proper acceleration. Does the string break? (It is light enough and strong enough not to break under its own weight.)

The problem can be extended in many interesting directions. Suppose the system is studied from other inertial frames. The events labelled by the initial firings of the two rocket engines are spacelike separated, so inertial frames exist in which A fires before B and others in which B fires before A. The elastic properties of the string will be important for a complete solution. It is also interesting to forget the string and to equip both ships with radar: use the radar echoes to measure the separation between the ships, in the instantaneous rest frame of one or the other .... Have fun.]

Lecture 8 THINGS THAT GO FASTER THAN LIGHT

Anyone who has learnt relativity has usually been thoroughly indoctrinated with the idea that nothing goes faster than light. A great many velocities exceeding c are encountered in physics: these may cause confusion and distress. They are therefore worth discussing, both to alleviate these undesirable effects and to clarify the content of special relativity.

What does relativity have to say about a limiting velocity? Recall the Lorentz transformations:

$$x' = \frac{x - vt}{\sqrt{1 - v^2/c^2}}; \qquad t' = \frac{t - vx/c^2}{\sqrt{1 - v^2/c^2}}$$

x, v and t; x' and t' are all measurable quantitive, and must be represented by real numbers. If the measured relative velocity between two inertial frames exceeded c, then this condition would be violated. The velocity of one inertial frame measured in another must always be $\leq c$. Consistent with this requirement we found that the velocity of an accelerated particle approaches c asymptotically, for any finite proper acceleration, and that the inertial mass of a particle is given by $E/c^2 = \frac{m_0}{\sqrt{1-v^2/c^2}}$. A particle of non-zero rest mass must acquire infinite energy to reach the speed of light; the machinery is the increasing inertial mass. Finally we found that the different time ordering of two events with spacelike separation, obtaining in different inertial frames, leads to no logical paradoxes provided that signals cannot be propagated from one frame to another at a speed exceeding the velocity of light. These are the restrictions imposed by the special theory of relativity.

We may now look at some examples of velocities which exceed that of light.

EXAMPLE 1: The pulsar in the Crab nebula (a neutron star left behind after the supernova explosion seen in 1054 AD) is $\sim$ 5000 light years distant from earth. It emits a tight beam of radiofrequency electromagnetic radiation (and light) which sweeps across the earth every 33 ms. What is the velocity with which this beam passes over the surface of a sphere of radius 5000 light years centred on the pulsar? The answer is the obvious one,

$$2\pi \times 5000 \text{ light years}/33 \text{ ms} \simeq 9 \times 10^{23} \text{cm s}^{-1} = 3 \times 10^{13} \text{ c}.$$

There is no problem. Energy is travelling radially outward at $\sim c$ and we cannot interrupt the beam we detect in order to send a message to someone 10,000 light years away on the opposite side of the sphere. The photons in the next burst to reach him are already nearly there and are spacelike separated from earth. If you think of the rotating beam as being modulated to form a series of rapid pulses, at some instant of time in the pulsar rest frame the pulses lie on a tightly wound spiral, and with time this spiral expands radially outward at the speed of light.

Another form of the problem is as follows. Fire a burst of microwaves at Venus, swing the antenna around (which takes a few minutes, unless it is a phased array) and fire a second burst at Mars. Energy travels at speed c towards Venus and Mars, but the arrival of the signals labels two events which are spacelike separated — Venus cannot tell Mars that it has received a signal until long after the signal from earth reaches Mars. We can however define a velocity $D_{vm}/\Delta t$ where D_{vm} is the distance from Venus to Mars and Δt is the difference in arrival times: it could be infinite.

[PROBLEM: Can you write on the screen of a CRT faster than light?]

EXAMPLE 2: Scissors. A steel rule is clamped to the laboratory bench and a second steel rule slides over it, with a velocity $v(< c)$ perpendicular to its length. The two rules make a small angle α.

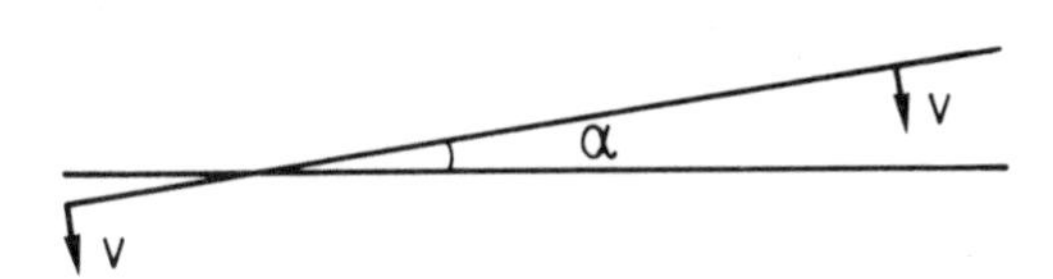

What is the speed with which the scissors point moves along the clamped rule?

The answer is the obvious one — $v/\sin\alpha$. This speed can exceed that of light for modest values of v if α is small enough, and for a guillotine may be infinite. Again there is no problem. Neither energy nor any material object is travelling faster than c, and it is impossible to send a message (such as — get out of the way) at the speed of the scissors point. Any interference at one particular intersection point would propagate through the steel at the speed of sound, $\ll c$. (The equation of state in the core of a neutron star is not well understood. An important constraint in the calculation of neutron star structure is that the velocity of sound must be $< c$).

EXAMPLE 3: An explosion takes place on a star at the centre of a planetary nebula, and a burst of X-rays streams outwards, causing the gas in the nebula to fluoresce.

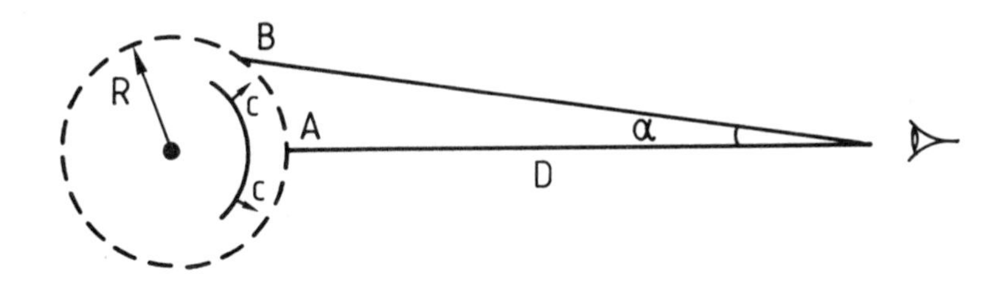

We first see a bright spot at A, which expands outwards. Fluorescent light from a bit of the nebula at B reaches us at a time $\sim D^2\alpha^2/2Rc$ after we first see the bright spot. The patch of fluorescence spreads at a perceived velocity $2cR/D\alpha \gg c$. In fact both A and B are switched on simultaneously. Neither energy nor material particles are travelling faster than light, and the events A turns on, B turns on are spacelike separated.

[PROBLEM: A number of very distant radio sources appear to be expanding tranverse to the line of sight with velocities in excess of $5c$. The best evidence comes from VLBI (Very Long Baseline Interferometry) observations of the quasar 3C 273, where a knot of radio emission was observed at a transverse distance of 62 light years from a central core in July 1977. By July 1980 the knot was 87 light years from the core: the observations give a constant transverse velocity for the knot of $9.6c$.

One possible model is that the knot was ejected from the core with high velocity $v_0 < c$, in a direction making a small angle ϕ with the line of sight. Show that the perceived transverse velocity v in such a model is given by

$$v = v_0 \sin\phi \Big/ \left(1 - \frac{v_0}{c}\cos\phi\right) .$$

Find $\phi_{\max}$, the maximum value of ϕ consistent with these observations. If all radio-emitting quasars produced two relativistic knotty jets of matter, what is the probability that a randomly chosen quasar will have a jet with $\phi < \phi_{\max}$? What is the minimum value of v_0 allowed by the observations, and the corresponding value of the angle ϕ ?

You might also like to consider the range of angles ϕ over which a relativistic jet (with velocity equal to the minimum value of v_0) might be observed.]

EXAMPLE 4: In a colliding beam experiment e^+ and e^- are circulating in opposite directions, each with energy 15 GeV. What is their relative velocity in the laboratory? The answer is marginally less than $2c$, the sum of the two laboratory velocities. But the velocity of e^+ in the e^- rest frame is still marginally less than c, and vice versa, and again there is no question of signals being propagated faster than light. But if you want to calculate a flux factor (in order to turn a transition rate into a cross section) in the laboratory frame, the relevant velocity is $\sim 2c$.

Most of these examples are straight forward, provided that they are analysed carefully, with the true restrictions of special relativity in mind. Somewhat more subtle are various aspects of phase velocity.

EXAMPLE 5: This one is trivial. A plane electromagnetic wave travels at velocity c in the direction normal to the wavefront (a plane of constant phase). Then the velocity with which a point of constant phase travels along a line making an angle α with the wavefront is $c/\sin\alpha$. This is the scissors problem again — neither energy nor a signal is propagating at this velocity. It helps clarify the next example.

EXAMPLE 6: In a rectangular waveguide, the complete solution for the electromagnetic field is a product of standing waves transverse to the guide and a propagating wave along the guide. Substitution of such a solution into the wave equation yields the condition

$$k_z^2 = \frac{\omega^2}{c^2} - \left(\frac{\pi\ell}{X}\right)^2 - \left(\frac{\pi m}{Y}\right)^2$$

where k_z is the propagation vector down the tube, ω the frequency, X, Y the transverse dimensions and ℓ, m are integers (which cannot be simultaneously zero). The phase velocity down the tube is $\omega/k_z > c$. Decompose the standing waves across the guide into a superposition of travelling waves. The following picture results:

Recalling Example 5, it is now obvious why the phase velocity down the tube exceeds c. Photons (and energy) are travelling obliquely at velocity c and so go down the tube with velocity $< c$. Direct calculation shows that both energy and signals (a disturbance of the monochromatic carrier) propagate down the tube at velocity $\frac{\partial \omega}{\partial k_z} < c$; $\frac{\partial \omega}{\partial k_z}$ is the definition of the group velocity down the tube.

If $\frac{\omega^2}{c^2} < \left(\frac{\pi \ell}{X}\right)^2 + \left(\frac{\pi m}{Y}\right)^2$ then k_z is imaginary and the fields fall off exponentially with distance down the tube, all points oscillating in phase, regardless of z, with frequency ω. The phase velocity is infinite — and no energy propagates.

EXAMPLE 7: In a dilute plasma the electrons are effectively free and it is trivial to calculate the refractive index, which is a function of frequency. For a plane wave the velocity is a function of frequency and the condition

$$k^2 = \frac{\omega^2}{c^2}\left\{1 - \frac{4\pi N e^2}{m_e \omega^2}\right\}$$

results, where there are N electrons cm^{-3}. The phase velocity is $\omega/k > c$.

It is worth noting that a plane monochromatic wave extends throughout the whole of space and for all time, and so is at best an approximation to a real situation. For such a mathematical construct, the phase velocity is the relation between ω and k. There is no relativistic objection to the phase velocity exceeding c provided energy is not propagated at a speed exceeding c, and that no signal can be propagated at a speed exceeding c. We may investigate the velocity at which energy is propagated while maintaining the fiction of a monochromatic plane wave. The energy flux is given by the Poynting vector and the velocity of energy propagation is obtained on dividing the flux by the energy density associated with the wave. This contains not only the energy stored in the fields and energy associated with polarisation, which are contained in the familiar expression $\frac{1}{8\pi}\{\mathbf{E}.\mathbf{D} + \mathbf{B}.\mathbf{H}\}$ but also the kinetic energy of the plasma electrons at the moment when $\mathbf{E} = \mathbf{0}$. This energy is there because the electrons are moving and the electrons are moving because of the wave. It would be put in there when the wave was first established (we then wait a very long time for transients to die out.) The velocity of energy propagation, as defined above, is $kc^2/\omega < c$.

An infinite plane monochromatic wave carries no signal. A signal may be imposed on this carrier wave, but then it is no longer monochromatic, and in such an ideal simple plasma the disturbance travels with the group velocity

$$\frac{\partial \omega}{\partial k} = kc^2/\omega < c.$$

Thus energy and signals propagate at the group velocity, $< c$. There is no contradiction with relativity. [When electrons are bound in atoms, there are situations where both ω/k and $\partial\omega/\partial k$ exceed c. In such a situation the mathematical construct $\partial\omega/\partial k$ is not an adequate approximation to signal velocity and loses physical significance. Neither energy nor signals propagate at speeds in excess of c. See L. Brillouin, Wave Propagation and Group Velocity (Academic Press 1960) or J.D. Jackson, Classical Electrodynamics (2nd Ed., Wiley 1975).] In a simple plasma, the phase velocity does exceed that of light. Metallic mirrors will reflect X-rays by total internal reflection at small glancing angles, because the refractive index of the metal is < 1. Near the sun, the plasma density falls off rapidly with distance. Pulses of radiofrequency from a pulsar passing close to the limb of the sun are deflected away from the sun (because the refractive index increases with increasing distance) but are delayed (because the group velocity is $< c$).

EXAMPLE 8: This one is wholly spurious. In a material medium, under most circumstances both the phase and group velocities of light are $< c$. A particle may exceed the velocity of light in a material medium, giving rise to Cerenkov radiation. There is no problem, for the limiting velocity is c, the velocity of light in vacuum, not the local velocity of light in a material medium.

Two speculative EXAMPLES: Are there things of physical significance which travel faster than light?

One interesting possibility is concerned with the structure of quantum mechanics. The polarisation correlations between two photons emitted successively from an atom are correctly predicted by quantum mechanics (the best current experiments are those of A. Aspect *et al. Phys. Rev. Lett.* **49** 91, 1804 (1982)) but violate limits set by any realistic local theory (Bell's inequalities, see J.S. Bell *Physics* **1** 195 (1965), reprinted in Quantum Theory and Measurement, Ed. J.A. Wheeler and W.H. Turek (Princeton 1983); B.D'Espagnat Scientific American Nov. 1979 128). Either we discard the assumption that photons really have a certain value of the polarisation, or something concerning the correlations propagates faster than light, for the two polarisation analysers receive the photons with space-like separation. (Signals cannot be sent this way — only the correlation is established and so to code information the measured polarisations at both stations are needed. The polarisation observed at one analyser can only be transmitted to the other at speeds $\leq c$). [The community of physicists remains unhappy about quantum mechanics, which seems to be right, but crazy.]

There has been speculation that although material particles cannot be boosted to speeds in excess of c, there might exist particles whose velocity is always in excess of c, from the moment of creation onwards. If such particles had no interaction whatever with the subluminal world, there would be no problem with relativity, for they would not exist. But suppose that such particles (tachyons) exist and do interact with the world we live in. (There is no experimental evidence whatever for the existence of tachyons). The idea is to keep the familiar kinematic relations of special relativity,

$$\begin{aligned} E &= \frac{m_0}{\sqrt{1-u^2}} \qquad & E^2 - p^2 &= m_0^2 \qquad \text{(units } c = 1) \\ \mathbf{p} &= \frac{m_0 \mathbf{u}}{\sqrt{1-u^2}} \qquad & \mathbf{p}/E &= \mathbf{u} \end{aligned} \tag{8.1}$$

to keep Lorentz covariance, and form a consistent interpretation for a class of particles for which $u \geq 1$ always, and yet E and $\mathbf{p}$ are real physical quantities. This requires that the proper mass m_0 is imaginary. This would be acceptable: we cannot boost into the rest frame of such a particle. The inertial mass is $m_0/\sqrt{1-u^2}$ and is real. Remember that u is obtained, in an ordinary inertial frame, from successive measurements of the position of the tachyon.

We can investigate the Lorentz covariance of these kinematic relations in various ways. First, because a tachyon travels faster than light in some inertial frame, two events linked by a tachyon are space-like separated. This spacelike interval is invariant under Lorentz transformations between any two inertial frames, so in all inertial frames the separation of the two events is spacelike and so in all inertial frames the tachyon speed exceeds c. The tachyon concept is Lorentz covariant. This can of course be verified directly from the velocity transformations.

Transform E, $\mathbf{p}$ in (8.1), defined in an inertial frame S in which the tachyon speed is $u > c$ into a frame S' in which the tachyon speed is $u' > c$. In frame S (8.1) hold; then in S'

$$E' = \frac{E - vp_x}{\sqrt{1-v^2}} \qquad p_{x'} = \frac{p_x - vE}{\sqrt{1-v^2}} \qquad p_{y'} = p_y$$

or

$$E' = \frac{m_0(1 - vu_x)}{\sqrt{1-v^2}\sqrt{1-u^2}} \qquad p_{x'} = \frac{m_0(u_x - v)}{\sqrt{1-v^2}\sqrt{1-u^2}} \qquad p_{y'} = \frac{m_0 u_y}{\sqrt{1-u^2}} \tag{8.2}$$

We also require

$$E' = \frac{m_0}{\sqrt{1-u'^2}} \qquad p_{x'} = \frac{m_0 u'_x}{\sqrt{1-u'^2}} \qquad p_{y'} = \frac{m_0 u'_y}{\sqrt{1-u'^2}} \tag{8.3}$$

The velocity transformation equations yield the relations

$$1 - u'^2 = \frac{(1-u^2)(1-v^2)}{(1-vu_x)^2}; \qquad \frac{u'_y}{\sqrt{1-u'^2}} = \frac{u_y}{\sqrt{1-u^2}}$$

so (8.2) and (8.3) are consistent. Thus any processes involving tachyon absorbtion or scattering which are allowed by energy-momentum conservation will satisfy the covariant conservation laws. (There are some surprises: a tachyon in free space may radiate a massless particle, for example).

There is however a problem. For particles with velocity $< c$ the energy is always defined positive. If we take the positive root of $1-u^2$ in (8.1) we should take the positive root in (8.3). Yet (8.2) shows that if $vu_x > 1$ a positive energy tachyon in S becomes a negative energy tachyon in S'. The only solution is to admit both positive and negative roots of $1-u^2$ in all frames, and keep negative energy solutions. The negative energy solutions have been employed in attempts to alleviate a different problem: if two events can be connected by tachyons then causality problems are anticipated, for spacelike separated events may have different time ordering in different inertial frames.

Take two events in S, one at the origin, the other at (x,t) such that $x = ut$ where $u > c$. Then in S'

$$t' = \frac{t - vx}{\sqrt{1-v^2}} = \frac{t(1-vu)}{\sqrt{1-v^2}}$$

If $vu > 1$ then the time order of the two events is reversed in S'. This is precisely the condition for a positive energy tachyon in S to become a negative energy tachyon in S'. Thus if a positive energy tachyon in S is emitted at the origin and absorbed later at (x,t), in S' a negative energy tachyon is absorbed at (x',t') and later emitted from the origin. This may be reinterpreted as emission of a positive energy antitachyon at (x',t') which is later absorbed at the origin.

There is at first sight a flaw in this reinterpretation. In S the emitter at the origin loses energy and the receiver at x gains energy, while in S' it is the other way round, regardless of reinterpretation. The apparent flaw arises from thinking of emission of positive energy as necessarily involving an atom changing its quantum state downwards (or classically decreasing its rest mass) and of absorbtion of positive energy as an atom changing to a state of higher energy. The states could be labelled. But tachyon kinematics contain surprises: a moving atom can emit a tachyon without changing its state. The energy changes at the two ends, viewed from either frame, are not inconsistent with the label changes. [See G. Feinberg *Phys. Rev.* **159** 1089 (1967)]. In this example causality is saved, in the sense that reinterpretation implies emission precedes absorbtion in all frames. But if you consider a modulated signal as opposed to a single tachyon, it is rather curious. Take successive points (x,t) such that $dx/dt = v$ and consider O in S playing the piano and transmitting to O' in S'. Reinterpretation implies that the tachyon communicator at O' is spontaneously radiating a Beethoven Sonata.... The only way out, if this is unacceptable, is for O to be unable to transmit a coded signal, as opposed to noise, via tachyons with $uv > 1$.

It gets worse. If meaningful signals can be transmitted by sufficiently high speed tachyon beams, then if O signals to O' and O' replies, the reply arrives before the message. You can't get out of this by arguing that both O and O' in fact sent two messages, because by employing a suitable chain of transponders reception of negative energy need not be involved at any station. [B. De Witt, *Physics Today*, December 1969, 49; F.A.E. Pirani, *Phys. Rev.* **D1** 3224, 1970.] If such tachyon messages can be sent, logical paradoxes arise (send a message to your parents instructing them to strangle you at birth.) The only way of preserving the causal structure of relativity would be for tachyonic apparatus to be incapable of transmitting messages under these circumstances.

Problems with causality also arise in considering the equation of motion of an electrically charged tachyon emitting Cerenkov radiation (in vacuo), losing energy and accelerating. Perhaps tachyons don't have electric charge? They must have a gravitational field and the same problems arise with gravitational Cerenkov radiation [F.C. Jones *Phys. Rev.* **D6** 2727 (1972)].

These problems did not (very properly) deter experimentalists from searching for tachyons (by looking for Cerenkov radiation in vacuum, searching for events in particle interactions in which a neutral particle was emitted with $|p| > E$, looking for precursor signals in cosmic ray showers). In the absence of either a theoretical or experimental imperative, interest in tachyons has waned. These speculations profer a stimulating mental gymnasium, but at present nothing more.

If particles or fields propagating at velocities in excess of c ever are discovered, we would be entering a wholly new realm of physics. Whether or not in such circumstances Lorentz covariance could somehow be retained would be of secondary importance. It must be remembered that physics is an experimental science, and that special relativity is the conceptual framework appropriate to the physical world as we today perceive it.

[GENERAL REFERENCE on tachyons: O-M. Bilianuk and E.C.G. Sudarshan, *Physics Today*, May 1969, 43. See also the correspondence following p47 of *Physics Today*, December 1969.]

Appendix — MAXWELL'S EQUATIONS

The four equations of Maxwell, together with the Lorentz force law, embody almost all of classical electromagnetism. We are not interested here in the average properties of fields in matter. It is advantageous (in connection with relativity) to work in units such that c appears as a fundamental constant, and my preference is to leave factors of 4π where they have a clear geometrical significance. The units employed here are the unrationalised gaussian system.

First, Coulomb's law in electrostatics. The force between two charges is proportional to the product of the charges and inversely proportional to the square of their separation, or

$$\mathbf{F}_2 = k\frac{q_1 q_2}{r_{12}^3}\mathbf{r}_{12} \qquad \text{where } k \text{ is a constant}$$

Define

$$\mathbf{E}(\mathbf{r}) = \frac{q\mathbf{r}}{r^3} \qquad \text{choosing } k = 1 \text{ (electrostatic unit of charge)}$$

Form $\int_S \mathbf{E}.\mathbf{d}S$ over a closed surface enclosing a point charge q. Since $\mathbf{r}.\mathbf{d}S/r^3$ is an element of solid angle

$$\int_S \mathbf{E}.\mathbf{d}S = 4\pi q$$

Generalise using linear superposition of electric fields

$$\int_S \mathbf{E}.\mathbf{d}S = 4\pi \sum_i q_i$$

where charges q_i are present inside the closed surface S. Let $\sum_i q_i \to \int \rho dV$ where ρ is the local charge density and obtain

$$\int_S \mathbf{E}.\mathbf{d}S = 4\pi \int_V \rho dV$$

where V is bounded by S. Apply Gauss' theorem in vector calculus and obtain the differential form of Coulomb's law

$$\nabla.\mathbf{E} = 4\pi\rho \qquad (A.1)$$

This is the first Maxwell equation.

Magnetic fields are only produced by circulating currents, as far as is known. The fundamental experimental result is the Biot-Savart Law for the force between two current carrying circuit elements

$$d\mathbf{F}_2 = k\frac{I_1 I_2 \mathbf{d}\boldsymbol{\ell}_2 \times (\mathbf{d}\boldsymbol{\ell}_1 \times \mathbf{r}_{12})}{r_{12}^3}$$

where $\mathbf{r}_{12}$ is the vector from 1 to 2 and k is a constant. If k is chosen equal to unity, the currents are measured in electromagnetic units. Define

$$d\mathbf{B}(\mathbf{r}) = \frac{I\mathbf{d}\boldsymbol{\ell} \times \mathbf{r}}{r^3}$$

If a current I is composed of charges streaming along, $I\mathbf{d}\boldsymbol{\ell} \to \rho\mathbf{v}dV$ and for a charged particle $\int \rho dV \to q$. The force on a charged particle in a magnetic field $\mathbf{B}$ is, in these units $\mathbf{F} = q\mathbf{v} \times \mathbf{B}$

The force between two charges is

$$F \sim \frac{q_{esu}^2}{r^2}$$

and between two currents

$$F \sim \frac{q_{emu}^2 v^2}{r^2}$$

The dimensions of charge in these two cases are different, and the ratio has the dimensions of a velocity. Define

$$q_{emu} = q_{esu}/c$$

The ratio q_{esu}/q_{emu} can be measured by comparing the rate of change of the force between two capacitor plates with the force between two circuit elements carrying the current generated by discharge of the capacitor. This is a quasi-static measurement: at this stage of development the equations do not contain electromagnetic radiation.

If all charges and currents are measured in *esu*, then

$$\mathbf{B}(r) = \frac{1}{c}\int \frac{I\mathrm{d}\boldsymbol{\ell} \times \mathbf{r}}{r^3} \qquad (A.2)$$

and the force acting on a charged particle in a general electromagnetic field is the Lorentz force

$$\mathbf{F} = q(\mathbf{E} + \frac{1}{c}\mathbf{v} \times \mathbf{B}). \qquad (A.3)$$

Vector algebra applied to (A.2) yields

$$\nabla.\mathbf{B} = \mathbf{0}$$

which is the second Maxwell equation and expresses the absence of (observed) magnetic monopoles. (A.2) also yields, under the assumption of steady currents

$$\nabla \times \mathbf{B} = \frac{4\pi}{c}\mathbf{J} \qquad (A.4)$$

where $\mathbf{J}$ is the local current density $\rho\mathbf{v}$.

The Faraday-Lenz law of electromagnetic induction is

$$\Sigma = -\frac{1}{c}\frac{d\Phi}{dt}$$

where Σ is the electromotive force, actually the integral of force on a unit charge round a circuit, and Φ is the magnetic flux threading that circuit, $\Phi = \int \mathbf{B}.\mathbf{d}s$. If part of the circuit moves, the resulting force is embodied in the Lorentz force law. For a fixed circuit, which need not involve real wires,

$$\Sigma = \oint \mathbf{E}.\mathrm{d}\boldsymbol{\ell} = -\frac{1}{c}\frac{\partial}{\partial t}\int \mathbf{B}.\mathrm{d}S$$

Applying Stokes' theorem

$$\nabla \times \mathbf{E} = -\frac{1}{c}\frac{\partial \mathbf{B}}{\partial t} \qquad (A.5)$$

which is the third Maxwell equation.

So far everything was derived more or less directly from experiment. Equation (A.4) is however unsatisfactory in the general case, for $\mathbf{J}$ is generally time dependent. The equation of continuity, which expresses local conservation of charge is obtained from the relation

$$\int_S \mathbf{J}.\mathbf{d}S = -\frac{\partial q}{\partial t}$$

where $\mathbf{J}$ is the current density flowing through the closed surface S which contains charge q. Application of Gauss' theorem yields

$$\nabla.\mathbf{J} + \frac{\partial \rho}{\partial t} = 0$$

Thus in general $\nabla.\mathbf{J} = -\frac{\partial \rho}{\partial t}$, yet taking the divergence of (A.4) yields

$$\nabla.(\nabla \times \mathbf{B}) = \frac{4\pi}{c}\nabla.\mathbf{J}$$

The left hand side is identically zero, the right hand side may be non-zero.

If, however, $\mathbf{J}$ is replaced in (A.4) by

$$\mathbf{J} \to \mathbf{J} + \frac{1}{4\pi}\frac{\partial \mathbf{E}}{\partial t}$$

$$\nabla.\left(\mathbf{J} + \frac{1}{4\pi}\frac{\partial \mathbf{E}}{\partial t}\right) = \nabla.\mathbf{J} + \frac{\partial \rho}{\partial t} \equiv 0$$

using (A.1). Maxwell's great contribution to electromagnetic theory was to postulate that the general differential equation relating $\mathbf{B}$ to its sources is

$$\nabla \times \mathbf{B} = \frac{4\pi}{c}\mathbf{J} + \frac{1}{c}\frac{\partial \mathbf{E}}{\partial t} \tag{A.5}$$

The complete set of Maxwell's equations now reads

$$\nabla.\mathbf{E} = 4\pi\rho \qquad \nabla.\mathbf{B} = 0$$

$$\nabla \times \mathbf{E} = -\frac{1}{c}\frac{\partial \mathbf{B}}{\partial t} \qquad \nabla \times \mathbf{B} = \frac{4\pi\mathbf{J}}{c} + \frac{1}{c}\frac{\partial \mathbf{E}}{\partial t}$$

In these equations, the sources ρ and $\mathbf{J}$ include polarisation charges and magnetisation currents, if present. To Maxwell's equations we add the equation of continuity and the Lorentz force law

$$\nabla.\mathbf{J} + \frac{\partial \rho}{\partial t} = 0$$

$$\mathbf{F} = q\left(\mathbf{E} + \frac{1}{c}\mathbf{v} \times \mathbf{B}\right)$$

Maxwell's equations admit plane wave solutions in empty space of the form

$$\mathbf{E} = \mathbf{E}_0 e^{i(\mathbf{k}.\mathbf{r} - \omega t)}$$

$$\mathbf{B} = \mathbf{B}_0 e^{i(\mathbf{k}.\mathbf{r} - \omega t)}$$

subject to the conditions

$$k^2 - \frac{\omega^2}{c^2} = 0$$
$$\mathbf{k}.\mathbf{E} = \mathbf{k}.\mathbf{B} = \mathbf{0}$$
$$\mathbf{k} \times \mathbf{E} = \frac{\omega}{c}\mathbf{B}$$

In radiation problems it is advantageous to work with the scalar and vector potentials: in quantum mechanics these are of greater significance than the fields $\mathbf{E}$, $\mathbf{B}$. They are constructed by first noting that $\nabla.(\nabla \times \mathbf{A}) \equiv 0$ so that $\nabla.\mathbf{B} = \mathbf{0}$ is automatically satisfied by the definition

$$\mathbf{B} = \nabla \times \mathbf{A}.$$

Then

$$\nabla \times \left(\mathbf{E} + \frac{1}{c}\frac{\partial A}{\partial t}\right) = 0$$

and since $\nabla \times (\nabla\phi) \equiv 0$

$$\mathbf{E} = -\nabla\phi - \frac{1}{c}\frac{\partial \mathbf{A}}{\partial t}$$

The two inhomogeneous equations now become

$$\nabla.\left(-\nabla\phi - \frac{1}{c}\frac{\partial \mathbf{A}}{\partial t}\right) = 4\pi\rho$$

$$\nabla \times (\nabla \times \mathbf{A}) = \frac{4\pi}{c}\mathbf{J} + \frac{1}{c}\frac{\partial}{\partial t}\left(-\nabla\phi - \frac{1}{c}\frac{\partial \mathbf{A}}{\partial t}\right)$$

These two equations may be separated by an appropriate choice of $\nabla.\mathbf{A}$

$$\nabla.\mathbf{A} + \frac{1}{c}\frac{\partial\phi}{\partial t} = 0$$

This is the Lorentz condition and corresponds to selecting the Lorentz gauge. This choice can be made because the physics is invariant under the substitution

$$\mathbf{A} \rightarrow \mathbf{A} + \nabla\chi$$
$$\phi \rightarrow \phi + \frac{\partial\chi}{\partial t}$$

— gauge invariance, a condition satisfied by Maxwell's equations which has assumed great importance in quantum field theory.

The decoupled equations for the scalar and vector potentials take the form

$$\nabla^2\phi - \frac{1}{c^2}\frac{\partial^2\phi}{\partial t^2} = -4\pi\rho$$
$$\nabla^2\mathbf{A} - \frac{1}{c^2}\frac{\partial^2\mathbf{A}}{\partial t^2} = -\frac{4\pi}{c}\mathbf{J}$$

The expressions for energy flux and energy density may be obtained from Maxwell's equations by noting that $\mathbf{J}.\mathbf{E}$ is the rate at which the fields perform mechanical work on charges. The substitution

$$\mathbf{J} = \frac{c}{4\pi}\left(\nabla \times \mathbf{B} - \frac{1}{c}\frac{\partial \mathbf{E}}{\partial t}\right) \qquad \text{yields}$$

$$\mathbf{J}.\mathbf{E} = -\frac{c}{4\pi}\nabla.(\mathbf{E}\times\mathbf{B}) - \frac{1}{8\pi}\frac{\partial}{\partial t}\left\{E^2 + B^2\right\}$$

leading to the identifications

$$\mathbf{S} = \frac{c}{4\pi}\mathbf{E}\times\mathbf{B} \qquad \text{for electromagnetic energy flux}$$

$$\mathcal{E} = \frac{1}{8\pi}\left\{E^2 + B^2\right\} \qquad \text{for electromagnetic field energy density.}$$

In empty space we have

$$\nabla.\mathbf{S} + \frac{\partial\mathcal{E}}{\partial t} = 0$$

which is an equation of continuity for electromagnetic field energy, expressing local conservation of energy.

INDEX

INDEX